Phosphorus-31 NMR Spectroscopy

Olaf Kühl

Phosphorus-31 NMR Spectroscopy

A Concise Introduction for the Synthetic
Organic and Organometallic Chemist

 Springer

Dr. Olaf Kühl
Universität Greifswald
Institut für Chemie und
Biochemie
Soldtmannstr. 23
17489 Greifswald
Germany
Email, personal: dockuhl@gmail.com

ISBN: 978-3-540-79117-1   e-ISBN: 978-3-540-79118-8

Library of Congress Control Number: 2008932569

Cover design: Künkel Lopka GmbH

Printed on acid-free paper

9 8 7 6 5 4 3 2 1

springer.com

Preface

Nuclear Magnetic Resonance is a powerful tool, especially for the identification of hitherto unknown organic compounds. ^{1}H- and ^{13}C-NMR spectroscopy is known and applied by virtually every synthetically working Organic Chemist. Consequently, the factors governing the differences in chemical shift values, based on chemical environment, bonding, temperature, solvent, pH, etc., are well understood, and specialty methods developed for almost every conceivable structural challenge. Proton and carbon NMR spectroscopy is part of most bachelors degree courses, with advanced methods integrated into masters degree and other graduate courses.

In view of this universal knowledge about proton and carbon NMR spectroscopy within the chemical community, it is remarkable that heteronuclear NMR is still looked upon as something of a curiosity. Admittedly, most organic compounds contain only nitrogen, oxygen, and sulfur atoms, as well as the obligatory hydrogen and carbon atoms, elements that have an unfavourable isotope distribution when it comes to NMR spectroscopy. Each of these three elements has a dominant isotope: ^{14}N (99.63% natural abundance), ^{16}O (99.76%), and ^{32}S (95.02%), with ^{16}O, ^{32}S, and ^{34}S (4.21%) NMR silent. ^{14}N has a nuclear moment $I = 1$ and a sizeable quadrupolar moment that makes the NMR signals usually very broad and difficult to analyse.

There are quite a few less common heteronuclei, particularly in Elementorganic Chemistry, with highly important applications in catalysis, C—C and C—N bond forming reactions, Medicinal Chemistry, Pharmacy, Green Chemistry and natural product synthesis, to name a few, that would make studying their NMR spectroscopy highly beneficial to that part of the chemical community that occupies itself with the research, production, and distribution of these chemicals.

In particular, ^{31}P (100%), ^{19}F (100%), ^{11}B (80.42%), and, to a lesser extent, ^{27}Al (100%), ^{29}Si (4.70%), and ^{195}Pt (33.8%) are arguably the most important heteronuclei in NMR spectroscopy. There are excellent books and reviews available that deal with some regions of the chemical shift range of these heteronuclei, together with a plethora of highly theoretical books and reviews on all aspects of instrumentation, algorithms, Hamiltonians, pulse sequences, etc., which may be very beneficial to the technician or the NMR specialist, but which are almost meaningless to the Synthetic Chemist. The Synthetic Chemist is interested in the identification of a compound, and thus uses the chemical shifts as a means to establish a link to the heteronucleus contained in the compound, and needs a means of identifying a chemical shift value that bears the structural aspects of his/her proposed compound.

Heteronuclear NMR is highly useful in this context, since a given compound normally contains only very few atoms of this nucleus, making the spectrum relatively simple, especially when compared to carbon or proton NMR.

Simplicity is needed in explaining the very complex field of phosphorus NMR to the non-specialist, and the Synthetic Chemist in particular. Simplicity is also the main shortcoming of this book; complex explanations are sometimes deliberately and necessarily oversimplified to keep the book in perspective and the intended reader in sight. I am far from apologetic in this regard, since I believe that it is better to teach 99 students to be right most of the time than just one to be completely right all of the time.

It is the primary aim of this book to enable the reader to identify the main factors governing the phosphorus chemical shift values in the ^{31}P-NMR spectrum, and to make an educated guess as to where the phosphorus resonance(s) of a given target compound can be expected. It is *not* within the scope of this book to enable one to predict a phosphorus chemical shift precisely, or even within a reasonable margin of error, with a few notable exceptions.

Whereas proton and carbon NMR spectroscopy is largely governed by σ-bonding contributions or well-defined π-bonded units, the influence of π-bonding interactions (hyperconjugation, negative hyperconjugation, and π-donor bonds) on the phosphorus chemical shifts is much more frequent and larger in magnitude. There are frequently no simple empirical formulae to describe the chemical environment of phosphorus atoms, making a quantitative calculation very complex and impractical. In fact, most theoretical computations of phosphorus chemical shifts take days, if not weeks, and plenty of financial and instrumental resources to produce the same (or worse) results as the educated guess of a seasoned researcher in the field.

The further intent of this book is to assist the reader in determining important issues, such as bond order, π-bonding contributions from substituents, the existence or non-existence of metallacycles, etc.; in short, to make structural assignments without the aid of X-ray crystal structure determinations or theoretical chemists, and to explain structural differences in solution and the solid state where appropriate.

I regret that the book requires a good knowledge of organometallic chemistry for those chapters dealing with phosphorus ligands and substituents bonded to metal atoms. Those whose research takes them into the realm of metal coordinated phosphorus compounds undoubtedly already possess this knowledge. For those who read on out of curiosity, my best advice to them is to peruse one of the many excellent textbooks available in that field.

Acknowledgments

I thank the University of Alabama for a Visiting Professorship

I especially thank Dr. Joseph Thrasher, former Head of the Chemistry Department, and Dr. Anthony J. Arduengo III., Saxon Professor of Chemistry, who worked together to appoint me to teach "Spectroscopic Methods in Organic Chemistry", the course that gave birth to this book.

My gratitude to the students who contributed their motivation, abilities, and valuable comments to the making of this book:

Aymara Albury
Dionicio Martinez Solario
Amanda Joy Mueller
Jane Holly Poplin
Jason Runyon
Craig Wilson

I want to thank Prof. Masaaki Yoshifuji and Dr. Christian Schiel for their support and encouragement, and Professors Pierre Braunstein (Université Louis Pascal, Strasbourg, France) and Masaaki Yoshifuji (Tohilku University, Sendai, Japan; University of Alabama) for their valuable comments on this manuscript.

I am indebted to Prof. Dietrich Gudat (Universität Stuttgart, Germany) for taking the time to give me solid advice on the manuscript from a NMR expert's perspective. The book would be much poorer without him, although I could not incorporate all his suggestions in deference to my less specialised readership.

Greifswald, June 2008   Olaf Kühl

Contents

List of Abbreviations

Ad	adamantyl
Bu^i	iso-butyl
Bu^n	*n*-butyl
Bu^t	*tert*-butyl
Bz	benzyl
$C^1_1 (4)$	from graph theory: a repetitive chain sequence consisting of four atoms, one of which is a hydrogen donor and another is a hydrogen acceptor
Cp	cyclopentadienyl
Cp^R	substituted cyclopentadienide
Cp^*	pentamethyl cyclopentadienide
Cp°	tetramethylethyl cyclopentadienide
Cy	cyclohexyl
d	doublet
δ	chemical shift
δ_C	carbon chemical shift
Δ_{coord}	coordination chemical shift
$\Delta\delta$	difference between two chemical shifts
ΔI	quantum transition of the nuclear spin
δ_p	phosphorus chemical shift
dppe	*bis*-diphenylphosphinoethane
dppm	*bis*-diphenylphosphinomethane
dppp	*bis*-diphenylphosphinopropane
Δ_R	ring chemical shift
E_{ar}	aryl effect parameter (QALE parameter)
Et	ethyl
fac	*fac* (Δ-shaped) substitution pattern on octahedron
FT	fourier transform
γ	magnetogyric ratio
gem	geminal
η	hapticity: number of atoms a ligand uses to bond to the same metal atom
HOMO	highest occupied molecular orbital
I	nuclear spin
I-effect	inductive effect

κ	number of atoms a ligand uses to coordinate to metal atoms
L	ligand, σ-donor
λ^3	trivalent (phosphorus)
λ^5	pentavalent (phosphorus)
LUMO	lowest unoccupied molecular orbital
μ	magnetic moment
	bridging atom or ligand
Me	methyl
M-effect	mesomeric effect
mer	*meridial* (T-shaped) substitution pattern on octahedron
Mes	mesityl, 2,4,6-trimethylphenyl
Mes*	supermesityl, 2,4,6-tris-*tert*-butylphenyl, also known as Smes
ν	frequency
OTf	triflate $CF_3SO_3^-$
PALP	phosphorus atom lone pair
Ph	phenyl
Pr^i	iso-propyl
Pr^n	*n*-propyl
p-Tol	*para*-tolyl
P_4	white phosphorus, a P_4-tetrahedron
q	quartet
QALE	quantitative analysis of ligand effects
π_p	π-acidity (QALE parameter)
s	singlet
σ^c	shift parameter
Σ	mathematical sign "sum of"
Smes	supermesityl, 2,4,6-tri-*tert*-butylphenyl; also known as Mes*
t	triplet
tbp	trigonal bipyramidal
thf	tetrahydrofuran
Θ	dihedral angle (coupling)
	Tolman's cone angle (QALE parameter)
Tol	tolyl
Tp	*tris*-pyrazolatoborate
triphos	*tris*-diphenylphosphino-*tert*-butane
VE	valence electrons
VSEPR	valence shell electron pair repulsion
VT	variable temperature
χ_d	σ-donor capacity (QALE parameter)
*	asymmetric atom
	chiral centre

Conventions

Stereochemistry

We will use the convention of Organic Chemistry to depict the stereochemistry around a given atom. That means that the central atom and all other atoms connected to it by a straight solid line are in the plane of the paper, an atom connected by a solid wedge protrudes out of the plane (is situated in front of the paper) and an atom connected to the central atom by a dotted line reclines behind the plane (is situated behind the paper). This is illustrated in the following figures:

We will normally use the view on the left hand side for square planar (octahedral) and tetrahedral geometries around the optically active atom. The second view for a tetrahedral geometry shows the numbering scheme according to the Cahn, Ingold, Prelog priority rules; whereas the third view shows that a tetrahedron consists of two pairs of substituents that are in planes perpendicular to each other.

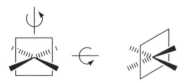

square planar

tetrahedral

NMR Standard

Phosphorus chemical shift values are given in ppm with 85 % H_3PO_4 as external standard (unless otherwise stated), with positive numbers indicating a downfield shift.

Downfield Chemical Shift

A downfield chemical shift is characterized by a deshielding of the nucleus, and thus an increase of the chemical shift value (toward the left of the spectrum).

Upfield Chemical Shift

An upfield chemical shift is characterized by a shielding of the nucleus, and thus a decrease of the chemical shift value (toward the right of the spectrum).

Chapter 1
Short Review of NMR Theory

This book does not intend to explain NMR spectroscopy in full. It does not even intend to explain enough theory to enable the reader to understand ^{31}P-NMR spectroscopy without prior knowledge of ^{1}H- or ^{13}C-NMR spectroscopy. There are many books that give a detailed explanation of ^{1}H- and ^{13}C-NMR spectroscopy. It is expected that the reader of this book is familiar with these nuclei in an NMR spectroscopic sense. However, a brief reminder of the basic concepts of NMR spectroscopy is given in this chapter as a way of introduction.

1.1 Origin of the Spectrum

An atomic nucleus contains protons, and, with the exception of ^{1}H, neutrons. Each proton and each neutron has an individual nuclear spin. The overall nuclear spin I of the nucleus is determined by vector addition of the individual proton and neutron spins.

Note: *Proton spins can only be canceled by proton spins and neutron spins can only be canceled by neutron spins.*

A quick estimate as to the nuclear spin being integer, half integer, or zero can be made from the number of protons (atomic number) and neutrons (atomic mass – atomic number) contained in that nucleus. A summary is given in Table 1.1.

A nucleus with a spin $I = 0$ has no NMR spectrum, it is NMR silent, and does not couple to other nuclei. Important examples are ^{12}C, ^{16}O, and ^{32}S. We are all familiar with the consequences from ^{13}C-NMR. Here the predominant isotope is ^{12}C, an NMR silent nucleus. The NMR active nucleus ^{13}C has a natural abundance of only 1.1%. Coupling to ^{1}H is observed and usually deactivated by broadband decoupling. Coupling to carbon is not observed, except in the case of ^{13}C enrichment, due to probability constraints.

Note: *Coupling to phosphorus – and other NMR active heteronuclei – is observed in ^{13}C-NMR spectroscopy.*

Nuclei with $I = 1/2$ are called dipolar nuclei. They appear to be spherical, with a uniform charge distribution over the entire surface. Since the nucleus appears to be

Table 1.1 A quick guide as to whether an isotope has zero, half-integer, or integer nuclear spin

I	Atomic Mass	Atomic Number	Nuclei
Half-integer	Odd	Odd	$^{1}_{1}H(\frac{1}{2})$, $^{31}_{15}P(\frac{1}{2})$
Half-integer	Odd	Even	$^{13}_{6}C(\frac{1}{2})$, $^{73}_{32}Ge(9/2)$
Integer	Even	Odd	$^{14}_{7}N(1)$, $^{2}_{1}H(1)$
Zero	Even	Even	$^{12}_{6}C(0)$, $^{16}_{8}O(0)$, $^{32}_{16}S(0)$

spherical, it disturbs a probing electromagnetic field independent of direction. The result is a strong, sharp NMR signal. ^{31}P is a dipolar nucleus.

Nuclei with $I > 1/2$ are called quadrupolar nuclei. They have a non-spherical charge distribution, and thus give rise to non-spherical electric and magnetic fields. They are said to have a quadruple moment. The consequences are major complications in their NMR spectra manifested as severe line broadening, shallow peaks, and difficulties with phasing and integration.

It is fortunate that ^{31}P is a dipolar nucleus. However, it is frequently bonded to quadrupolar nuclei, such as ^{14}N, ^{59}Co, ^{63}Cu, ^{65}Cu, ^{105}Pd, ^{193}Ir, and ^{197}Au. The magnitude of complications depends on factors such as natural abundance of the isotope, quadruple moment, and the relative receptivity.

Note: *Phosphorus atoms bonded to quadrupolar nuclei can experience severe line broadening.*

The standard NMR experiment for a dipolar nucleus like ^{31}P involves subjecting the probe to an external magnetic field with constant field strength $\mathbf{B}_0$ and variable radiofrequency v. In such a field, the phosphorus nuclei can either align themselves parallel or antiparallel to the external field. This can be represented as N_{α} or N_{β} (see Fig. 1.1). The nuclear magnetic moment μ is directly proportional to the spin I of the nucleus.

$$\mu = \gamma I\, h/2\,\pi$$

The proportionality constant γ is called the magnetogyric ratio and is specific to every nucleus. The energy ΔE required for a transition between the two states N_{α} and N_{β} is the transition energy

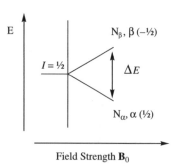

Fig. 1.1 A graphical description of the transition energy for a ^{31}P nucleus

$$\Delta E = (h\,\gamma / 2\,\pi)$$

Note: *Only single quantum transitions, $\Delta I = 1$, are allowed in NMR spectroscopy.*

As NMR data is usually expressed in terms of frequency, it is better to convert the transition energy ΔE to the corresponding frequency ν using the relationship $\Delta E = h\nu$.

$$\nu = (\gamma / 2\,\pi)\,\mathbf{B}_0$$

We know from ^{13}C and ^{1}H-NMR that there is a constant ratio between the resonance frequency and the applied field strength. The latter is an unchangeable parameter in modern NMR spectrometers.

$$\nu / \mathbf{B}_0 = \gamma / 2\,\pi$$

The proportionality constant is the magnetogyric ratio, which, for a given nucleus, in our case ^{31}P, is a constant ratio between the radiofrequency and the applied magnetic field. Since the applied magnetic field is constant for a given NMR spectrometer, we will need to change the radiofrequency ν, the MHz, for each nucleus that we wish to measure.

Note: *NMR spectrometers are named after the frequency they use to measure ^{1}H spectra. The required frequency for another nucleus, X, can be calculated from the magnetogyric ratio of that nucleus.*

$$\nu_X = \gamma_X \cdot \nu_H / \gamma_H$$

Note: *As the frequency of the NMR spectrometer is increased, the field strength of the magnet has to be increased in equal proportion. That is the reason for the infrastructural difficulties encountered with NMR spectrometers over 500 MHz.*

From Table 1.2, it can be seen that the magnetogyric ratio for phosphorus is about 2.5 times smaller than that for hydrogen. The required radiofrequency is therefore also about 2.5 times smaller than that of the spectrometer. For a 400 MHz NMR spectrometer, that would calculate to approximately 161 MHz.

So far, we have discussed the nucleus in terms of a specific isotope; in our case, ^{31}P. NMR spectroscopy is only useful to us because there is a difference between phosphorus atoms in different local environments. Each phosphorus atom in its own local environment is assigned its own chemical shift value, δ_p, measured in ppm. We know from ^{13}C- and ^{1}H-NMR that the origin of the chemical shift lies in the influence (shielding and deshielding) that the neighbouring atoms assert on the phosphorus atom. Mathematically, we can describe this phenomenon by the following equation:

$$\nu_{\text{eff}} = (\gamma / 2\,\pi)\,\mathbf{B}_0\,(1-\sigma)$$

Table 1.2 The magnetogyric ratio and NMR frequencies of common nuclei

Nucleus	Natural Abundance [%]	Magnetic Moment μ [μ_N]	Magnetogyric Ratio γ [10^7 rad T^{-1}s^{-1}]	NMR frequency Ξ [MHz]	Standard	Relative Receptivity D^P	Relative Receptivity D^C
^{1}H	99.985	4.83724	26.7519	100.000000	TMS	1.00	5.67×10^3
^{13}C	1.108	1.2166	6.7283	25.145004	TMS	1.76×10^{-4}	1.00
^{15}N	0.37	−0.4903	−2.712	10.136783	MeNO$_2$, [NO$_3$]$^-$	3.85×10^{-6}	2.19×10^{-2}
^{19}F	100	4.5532	25.181	94.094003	CCl$_3$F	0.834	4.73×10^3
^{29}Si	4.70	−0.96174	−5.3188	19.867184	TMS	3.69×10^{-4}	2.10
^{31}P	100	1.9602	10.841	40.480737	85% H$_3$PO$_4$	0.0665	3.77×10^2

The factor $(1-\sigma)$ converts the general resonance condition for the nucleus ^{31}P to the specific resonance condition for the individual phosphorus atom in the actual compound that we want to measure. The parameter σ carries all of the influences that our individual phosphorus atom is subject to. These are normally limited to diamagnetic factors, such as electronegativity of substituents, and diamagnetic anisotropy due to participation in double bonds or shielding cones of adjacent functional groups.

Note: *We will see that π-bonding has a very large influence on the chemical shift values for phosphorus atoms in ^{31}P-NMR spectroscopy.*

1.2 Coupling to other Nuclei

^{31}P couples to any other nucleus with a nuclear spin $I > 0$. Apart from ^{1}H and ^{13}C (the latter is only observable in carbon NMR spectra), ^{31}P nuclei couple to themselves (P—P coupling), fluorine (^{19}F), boron (^{10}B and ^{11}B), and a variety of metals, especially transition metals.

Most transition metals have a rather low natural abundance of the NMR-active isotope. The appearance of the ^{31}P-NMR spectrum is therefore that of a strong singlet flanked by satellite peaks of the respective multiplicity (see Fig. 1.2).

Note: *If the multiplicity is odd, then the strong singlet in the centre coincides with the central line of the multiplet. Thus, the central line of the multiplet is much enlarged.*

Note: *If the metal has more than one NMR active isotope, more than one satellite multiplet will be observed.*

Coupling to ^{1}H can be deactivated using proton broadband decoupling techniques. The respective notations are ^{31}P-NMR for the proton coupled spectrum, and ^{31}P-{^{1}H}-NMR for the proton decoupled spectrum. This notation is analogous to that used in ^{13}C-NMR spectroscopy.

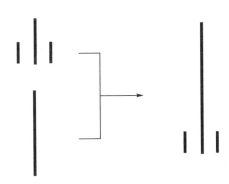

Fig. 1.2 A triplet superimposed on a singlet (looks like a doublet satellite)

Bibliography

Grim S O, Mitchell J D, Inorg Chem 16 (1977) 1770.

Harris R K, Mann B E, NMR and the Periodic Table, Academic Press 1978.

Kemp W, NMR in Chemistry, A Multinuclear Introduction, McMillan, 1986.

Nelson J H, Nuclear Magnetic Resonance Spectroscopy, Prentice Hall, Upper Saddle River, NJ (USA) 2003.

Pregosin P S, Transition Metal Nuclear Magnetic Resonance Studies in: Inorganic Chemistry Vol. 13, Elsevier, 1991.

Pregosin P S, Kunz R W, ^{31}P and ^{13}C NMR of Transition Metal Phosphane Complexes, Springer-Verlag, Heidelberg, 1979.

Silverstein R M, Webster F X, Kiemle D J, Spectrometric Identification of Organic Compounds, 7th Edition, John Wiley & Sons, Inc, New York, 2005.

Yadav L D S, Organic Spectroscopy, Kluiwer Academic Publishers, Dordrecht (NL) 2005.

Chapter 2
The Range of Chemical Shifts, Coupling Constants, and What Influences Each

For over a century, since Lewis introduced his octet rule, chemists have been accustomed to seeing the world of molecules through the glasses of two electron two center bonds, commonly represented by a straight line between two atoms. This image, developed into the valence bond VB-theory, is so powerful that it was frequently and craftily adopted to accommodate every increase of insight into the bonding between atoms in molecules achieved by progress in theoretical chemistry methods, or in rival concepts such as the molecular orbital theory. We now know of two electron three center bonds, four electron three center bonds, one electron bonds, and many more, and we have a VB representation for each.

We have also developed a rich canon of names to describe our rapidly changing perception of bonding in molecules, among which the terms "hyperconjugation", "negative (anionic) hyperconjugation", "Heitler-London (HL) Increased-Valence-Structures," and "π-Donor Bonds" are those that concern us. We use them to explain upfield and downfield chemical shifts in the ^{31}P-NMR spectra of compounds that we could not explain by "normal" σ-bonds alone.

The common characteristic behind all of the above terms is the occupation of previously empty orbitals with π-symmetry by electrons otherwise occupying orbitals with σ-symmetry. The term "π-bonding interaction," frequently used in this book, denotes a situation in which an (antibonding) orbital of π-symmetry receives electron density from phosphorus or a neighbouring atom, thus significantly influencing the phosphorus chemical shift. Nothing is implied as to the actual origin of this electron density. It suffices for the main purpose of this book, which is to predict the approximate region of the phosphorus resonances of a compound to aid the synthetic chemist in identifying (new) compounds and describing their ^{31}P-NMR spectra.

However, we do need a basic understanding of the principal causes for these π-bonding interactions. To this effect, brief and simple descriptions in lieu of a detailed explanation are given for three main terms found in the literature.

π-**Donor Bond** The ability of a substituent to use a fully occupied p-orbital to interact with the π-orbital of a P=X bond or an empty p-orbital of phosphorus.

These π-donor bonds require coplanarity of the participating groups in order to effectively form the four electron three center bond.

O. Kühl, *Phosphorus-31 NMR Spectroscopy*,
© Springer-Verlag Berlin Heidelberg 2008

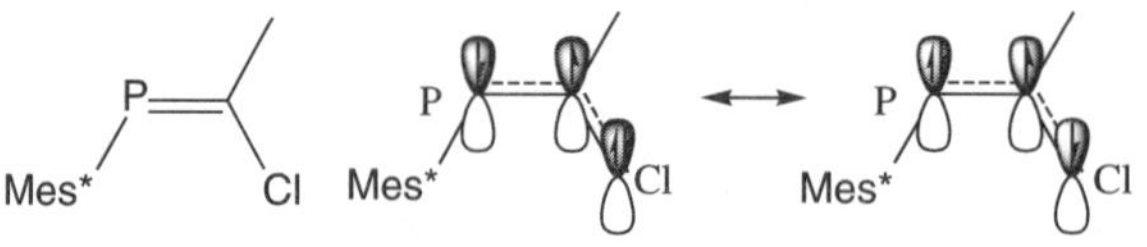

Fig. 2.1 Graphical representation of π-donor bonding in a color-phosphaalkene

The case, of course, becomes clearer when the phosphorus atom carries a positive charge. Then the familiar situation of a full p-orbital on the X atom interacting with an empty p-orbital on P to form P=X$^+$ arises.

Hyperconjugation Hyperconjugation originally describes the conjugation of a σ-bond with a double or triple bond. It is, therefore the interaction of the π-symmetric orbitals of this single bond with a p-orbital of this double or triple bond. More generally, it is a $\sigma \Rightarrow \pi^*$ or $n \Rightarrow \pi^*$ interaction. In Fig. 2.2, we see an interaction that results in a partial double bond character of the P-N bond.

The main difference between hyperconjugation and π-donor bonds, in practical terms, is that hyperconjugation does not require coplanarity, but is effective for all angles $\neq 90°$. The magnitude of the effect is angle dependent.

Negative Hyperconjugation Negative hyperconjugation, also known as anionic hyperconjugation, is essentially the complementary interaction to hyperconjugation, with opposite electron flow. Consequently, we have to denote it $\pi \Rightarrow \sigma^*$ or $\pi \Rightarrow n$.

The complex bonding situation in phosphaneoxides (-sulfides and -imines) can be simplified to a description of negative hyperconjugation, where the double bond resolves into a lone pair (on E) and a positive charge, located on phosphorus. Not surprisingly, with reference to the double bond rule, the third row element sulfur prefers charge separation, whereas the second row elements oxygen and nitrogen preserve a larger double bond character.

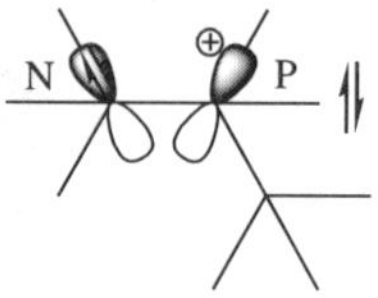

Fig. 2.2 Hyperconjugation
in an N-P$^+$ cation

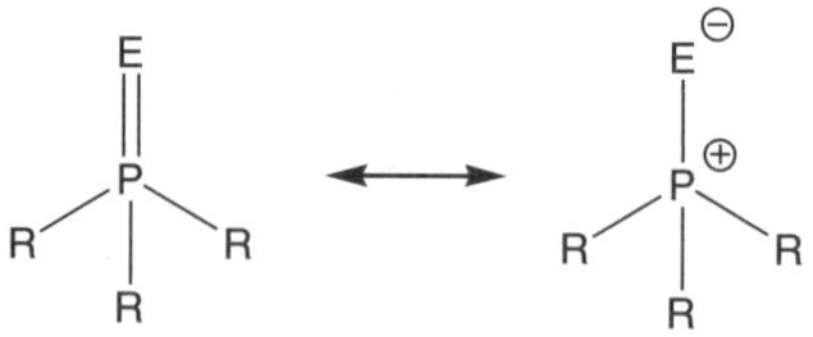

Fig. 2.3 The bonding in
phosphaneoxides, -sulphides,
and -imines

E = NR, O, S

2.1 Chemical Shift Values

The range of ^{31}P chemical shifts in diamagnetic compounds covers some 2000 ppm, and is thus one order of magnitude larger than that of carbon, and two orders of magnitude larger than proton. The upfield end of the chemical shift range in ^{31}P-NMR spectroscopy is defined by white phosphorus P_4 at $\delta_P = -527$ to -488 ppm, depending on solvent and water content of the sample. The downfield end of the range has been rather dynamic in recent decades, with $[\{Cr(CO)_5\}_2(\mu-PBu^t)]$ featuring a phosphorus resonance at $\delta_P = 1362$ ppm.

As the example of white phosphorus P_4 shows, ^{31}P chemical shifts are sensitive to solvents and to the presence of other compounds. They are also sensitive to temperature (and pressure). These influences will not be discussed in this book, but the reader should bear them in mind when looking at ^{31}P-NMR data or planning an experiment.

Today, the accepted reference standard for ^{31}P-NMR spectroscopy is 85% H_3PO_4 (external), meaning that a sealed ampule containing 85% aqueous H_3PO_4 submerged in a NMR tube filled with D_2O is measured and stored on the NMR spectrometer's hard drive as a reference. Phosphorus NMR does not require the use of deuterated solvents, but today's NMR spectrometers need a few drops of deuterated solvent for the locking signal. Although NMR measurements can be done without lock, this is not recommended for the average user (it is possible for a motorbike to fly over 24 parked busses. However, most people wisely abstain from the attempt).

Back in the 1960s, a dependance of the phosphorus resonance on the bond angle, electronegativity of the substituents, and the π-bonding character of the substituents was formulated. These three factors can be explained as: degree of s-orbital character in the bonds formed by phosphorus (bond angle), electron density on phosphorus (electronegativity), and the shielding cones of unsaturated systems. The shielding cones are a well known phenomenon in ^{1}H-NMR spectroscopy, and can be applied in ^{1}H- and ^{31}P-NMR.

Going one step further, we might ask about the influence that a phosphorus multiple bond may have on the chemical shift. From ^{13}C-NMR, we know that an olefin resonates downfield from an alkane, and an acetylene is found in between, but closer to the alkane. This is explained by diamagnetic anisotropy and the behaviour of P—C, P=C, and P≡C bonds should be analogous.

In Fig. 2.4 we can follow the trend throughout the phosphorus chemical shift range. At the upfield end, white phosphorus P_4 resonates at $\delta_P = -488$ ppm. This is readily explained by an electron rich cluster structure with P-P-P angles of $60°$. Coordination to a transition metal causes a reduction in electron density on phosphorus, as a P_4 lone pair is used to create a donor bond.[1] Similarly, substitution of a phosphorus atom by an isolobal transition metal fragment causes a downfield shift, as electron density is transferred from the electron rich phosphorus atoms to the "poorer" transition metal.

---

[1] In general, negative coordination shifts are possible. A description of coordination shifts and their causes is given in Chap. 7.

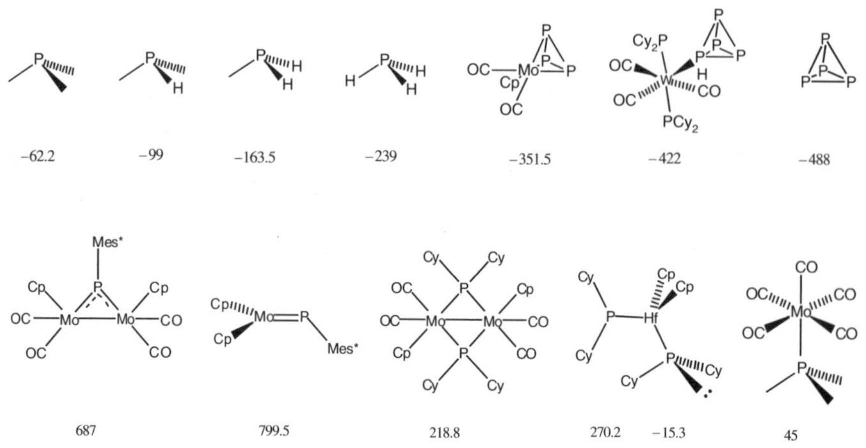

Fig. 2.4 The ^{31}P chemical shift range for phosphorus

The difference between the P_4 cluster and a phosphane is manifested in going to the monophosphane PH_3, a downfield shift of $\Delta\delta = 249$ ppm. In this transition, we see not only electronic changes, but also bond angle changes as the P_4 cluster is broken down. Substituting hydrogen for the more electronegative carbon in the phosphanes results in the expected downfield shift in line with the change of electronegativity.

From ^{13}C– and ^{1}H–NMR spectroscopy, we know that alkyl substitution on phenyl rings results in an upfield shift of the remaining C–H resonances. Phosphorus chemical shifts behave differently. Alkyl substituents have a – I effect on phosphorus. The phosphorus resonance is shifted downfield upon increasing alkyl substitution (see Table 2.1). Phenyl substituents cause an upfield shift compared to respective alkylated phosphorus compounds. PPh_3 has a chemical shift of $\delta_P = -6$ ppm, whereas PCy_3 has one of $\delta_P = 9$ ppm. In addition, phosphorus chemical shifts are influenced by changes in the Tolman cone angle.

A summary of phosphorus chemical shifts for phosphanes is given in Table 2.2, and a graphical display in Fig. 2.5. Both clearly illustrate the downfield shift of the

Table 2.1 The effect of substituents on the ^{31}P chemical shift of phosphorus

Molecule	δ_P [ppm]	Molecule	δ_P [ppm]	Molecule	δ_P [ppm]
PMe_3	−62.2	PBu^t_3	63	PPh_3	−6
PMe_2Cl	92	PBu^t_2Cl	145	PPh_2Cl	81.5
$PMeCl_2$	191.2	PBu^tCl_2	198.6	$PPhCl_2$	165
PCl_3	218				

Table 2.2 The dependence of phosphane phosphorus chemical shifts on the number of hydrogen substituents

Compound Class	Formula	δ_P [ppm]
Primary phosphanes	CPH_2	−170 to −70
Secondary phosphanes	C_2PH	−100 to 20
Tertiary phosphanes	C_3P	−70 to 70

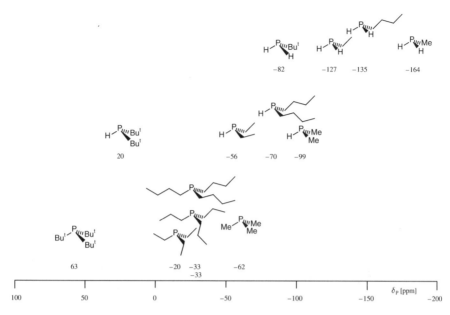

Fig. 2.5 A graphical visualization of the dependance of phosphine phosphorus chemical shifts on the number of hydrogen substituents

phosphorus resonance upon substituting hydrogen with carbon as one progresses from primary phosphanes via secondary phosphanes to tertiary phosphanes.

In Fig. 2.5, we see the overall trend that substitution of a carbon substituent with hydrogen results in an upfield chemical shift. The phosphanes move from the bottom left to the top right corner of the graph. Closer inspection shows a few exceptions in the overall trend as we examine the respective alkyl substituents in a given series.

In the series of tertiary phosphanes PMe_3, PEt_3, PPr^n_3, PBu^n_3, and PBu^t_3, we would expect a continuous downfield shift as the length of the alkyl chain increases. The last phosphane in the series, PBu^t_3, should show an additional contribution for steric effect. The actual chemical shift values are $\delta_p=-62$ ppm, $\delta_p=-20$ ppm, $\delta_p=-33$ ppm, $\delta_p=-33$ ppm, and $\delta_p=63$ ppm, meaning that from going to ethyl to n-propyl, the chemical shift difference is $\Delta\delta=-13$ ppm upfield, and not downfield as expected. Similarly, there is no chemical shift difference between n-propyl and n-butyl ($\Delta\delta=0$ ppm). The final entry, PBu^t_3, is $\Delta\delta=83$ ppm downfield from PEt_3, and thus in the expected region, if one takes the observed chemical shift difference between PMe_3 and PEt_3 ($\Delta\delta=42$ ppm downfield) as representative.

We know from ^{13}C-NMR spectroscopy that substitution with an electronegative functional group causes a diminishing downfield chemical shift for the carbon atoms along the alkyl chain (see Fig. 2.6). This pattern is interrupted for the γ-carbon, which experiences a typical upfield shift instead. If we apply this knowledge to our alkyl phosphanes, we will expect that there should be an upfield shift due to γ-substitution for PPr^n_3 and PBu^n_3, but not for PBu^t_3, as the latter has *tert*-butyl substituents for which the phosphorus atom is ß and not γ.

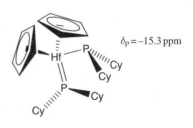

ε   γ   α
22.8   25.8   61.9 δ_C [ppm]

OH

14.2   32.0   32.8 δ_C [ppm]
ω   δ   β

Fig. 2.6 Illustration of the γ-effect in ^{13}C-NMR spectroscopy

Coordinating the phosphane PMe_3 to a transition metal again results in the expected downfield shift of $\Delta\delta = 107$ ppm, since the phosphorus lone pair is utilized for the donor bond, resulting in a decrease of electron density on phosphorus. However, when we form covalent bonds between phosphorus and the metal, the picture changes. Looking at $[Cp_2Hf(PCy_2)_2]$, we see two different phosphorus resonances at $\delta_p = 270.2$ ppm and $\delta_p = -15.3$ ppm, respectively (see Fig. 2.7). We note that the two phosphorus atoms are trigonal planar and tetrahedrally coordinated, and suspect that they are part of a P=Hf double and a P-Hf single bond, respectively.

Note: *P=M double bonds resonate downfield from P-M single bonds.*

The appearance of a bridging phosphide is seen downfield from a terminal one, and compounds with well defined P=M double bonds exhibit chemical shift values that are very significantly downfield. An example is $[Cp_2Mo=PMes^*]$ at $\delta_p = 799.5$ ppm.

The influence of P=C double bonds and P≡C triple bonds is shown in Fig. 2.8. It is easily seen that the region of P≡C triple bonds connects at the downfield end of tertiary phosphanes, and is in the same region as that of the heteroallene P=C=C system. The two examples feature chemical shifts of $\delta_p = 34.4$ ppm and $\delta_p = 70.6$ ppm, respectively.

$\delta_P = -15.3$ ppm

$\delta_P = 270.2$ ppm

Fig. 2.7 The two ^{31}P-NMR signals in $[Cp_2Hf(PCy_2)_2]$

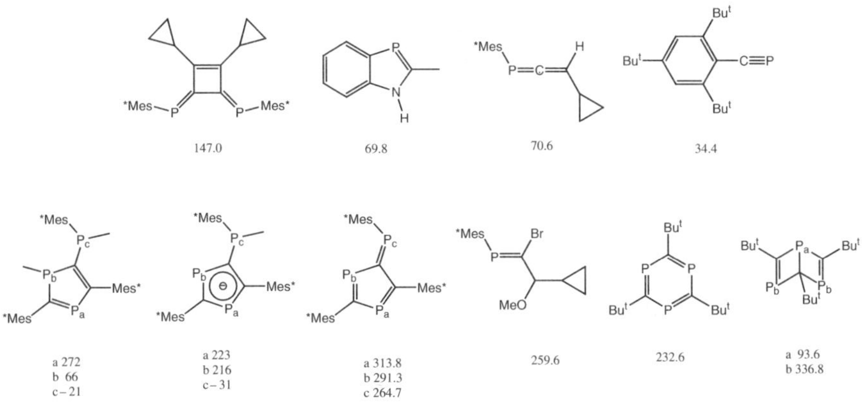

Fig. 2.8 ^{31}P-NMR chemical shift values for phosphorus carbon multiple bonds

Organic chemistry tells us that five-membered heteroaromatic ring systems are electron rich, and the corresponding six-membered ones are electron poor. This is corroborated by the ^{13}C-NMR spectra of pyrrole (five-membered N-heterocycle) and pyridine (six-membered N-heterocycle). The former has carbon chemical shifts of $\delta_p = 118.5$ and 108.2 ppm upfield from those of the latter, at $\delta_p = 150.6$, 124.5 and 136.4 ppm. With ^{31}P-NMR, we can measure the corresponding P-heterocycles more directly. The 1,3,5-triphosphabenzene resonates at $\delta_p = 232.6$ ppm, considerably downfield from the benzazaphosphole at $\delta_p = 69.8$ ppm, as expected. Interestingly, the 3,4 diphosphinidene-cyclobutene, featuring a conjugated, non-aromatic π-electron system with resonance structures, has a phosphorus chemical shift of $\delta_p = 147.0$ ppm midway between the two.

The Dewar analog to 1,3,5-triphosphabenzene is stable, and shows phosphorus resonances at $\delta_p = 93.6$ and 336.8 ppm. These ^{31}P chemical shifts are evidence for the folded roof-like structure of the compound, featuring a tertiary phosphane phosphorus atom at $\delta_p = 93.6$ ppm, and an isolated P=C double bond at $\delta_p = 336.8$ ppm, respectively.

In ^{13}C-NMR, we can compare the carbon chemical shift of benzene at $\delta_p = 128$ ppm with that of the cyclopentadienide ligand Cp$^-$ at $\delta_p = 118$–136 ppm, depending on the nature of the cationic metal fragment that it binds to. In ^{31}P-NMR, we can make this comparison between 1,3,5-triphosphabenzene with a phosphorus resonance at $\delta_p = 232.6$ ppm, and the 1,3,6 triphosphafulvenide anion with the relevant phosphorus atoms resonating at $\delta_p = 216$ and 223 ppm, respectively.

Note: *Similar chemical environments produce similar chemical shifts.*

Note: *Phosphorus chemical shifts are predictable.*

Having seen that phosphorus chemical shifts are indeed predictable in terms of electronegativity of substituents and π-bond involvement of phosphorus, we will now turn to the question of bonding to other moieties. We clearly expect a downfield shift, since the electron density on phosphorus will be reduced upon utilization of the lone pair for a donor bond.

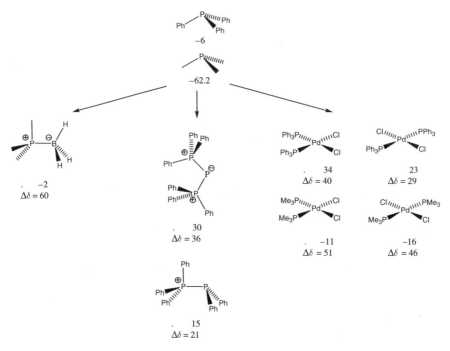

Fig. 2.9 The influence of coordination to a Lewis acid on the ^{31}P chemical shift

Note: *For valid comparisons, the other parameters have to be kept approximately constant, i.e. bond angles, cone angles, substituents (when bonding to metals is examined).*

In Fig. 2.9, we see a moderate downfield shift for the formation of a donor bond upon coordination, almost irrespective of the Lewis acid involved. It is the almost that interests us greatly. We use two phosphanes for comparison, PMe$_3$ and PPh$_3$, and note that the downfield shift for PPh$_3$ upon coordination ranges from $\Delta\delta = 21$–40 ppm, and that for PMe$_3$ is in an even narrower band of $\Delta\delta = 46$–60 ppm. In each case, the actual downfield shift depends on the Lewis base. Defining factors are the geometry around the metal (*cis* or *trans*), and the substituents on phosphorus (P$^+$ or PPh$_2^+$). Both factors are explainable and will be treated in the respective chapters.

Note: *PF$_3$ has a chemical shift value of $\delta_P = 97$ ppm, upfield from the range stated in Table 2.3 above.*

Table 2.3 The ^{31}P chemical shift regions of phosphorus halides

Formula	δ_P [ppm]
PHal$_3$	120–225
CPHal$_2$	155–190
C$_2$PHal	60–150

Table 2.4 The ^{31}P chemical shift values for selected phosphorus trihalides

Phosphorus Trihalide	PCl_3	PCl_2F	$PClF_2$	PF_3
δ_P [ppm]	220	224	176	97

Table 2.5 The influence of the halogen on the phosphorus resonance of the phosphorus trihalide

	PI_3	PBr_3	PCl_3	PF_3	$P(CN)_3$
δ_P [ppm]	178	227	218	97	−136
		$O=PBr_3$	$O=PCl_3$	$O=PF_3$	
δ_P [ppm]		−102.9	2	−35.5	

By far more interesting are the phosphorus chemical shifts that cannot be explained by the simple concepts of electronegativity, π-bonding, and coordination alone. We will first reexamine the evidence for alkyl phosphanes and for the respective halides. In Table 2.2, we see a perfect dependance of the phosphorus chemical shift upon the change in electronegativity as we go from primary phosphane to secondary phosphane to tertiary phosphane. This seems to be mirrored by the halides, as we inspect Table 2.3. However, compilations can be deceptive and PF_3 does resonate at $\delta_P=97$ ppm, in the range of C_2PHal in Table 2.3.

Closer inspection of the data for phosphorus trihalides shows us that there seemingly is a clear dependance on the nature of the halogen. We expected to see a correlation with the electronegativity of the halogen, meaning a downfield shift as we progress up the group from iodine to fluorine. In reality, however, PF_3 resonates at $\delta_P=97$ ppm, a startling $\Delta\delta=-123$ ppm upfield from PCl_3 at $\delta_P=220$ ppm.

We know from the chemistry of halogenated aromatic ring systems that fluorine can backbond to the aromatic π-electron system using one of its lone pairs. This ability is far less pronounced in chlorine, and entirely absent in the higher homologues bromine and iodine. The data in Table 2.5 corroborates this assumption for the interaction between phosphorus and halides.

Note: *Substitution of chlorine for fluorine essentially results in an upfield shift, despite fluorine being more electronegative than chlorine.*

The explanation is a π-bonding interaction (hyperconjugation) between the fluorine lone pair and phosphorus, resulting in an increase of electron density on phosphorus, and thus an upfield shift in the phosphorus resonance.

Note: *Substituents capable of π-bonding interactions cause additional shifts: upfield for a donor interaction, and downfield for an acceptor interaction.*

2.2 Coupling Constants

The coupling constants in ^{31}P-NMR spectroscopy are generally larger in magnitude than those in ^{1}H- or ^{13}C-NMR spectroscopy, but they are governed by the same principles. ^{1}J coupling constants of over 1000 Hz are readily observed. Coupling is facilitated through the σ-bonds of the backbone. There is no pronounced π-effect as the one observed for the chemical shift values, but there are examples for which a limited π-effect is indeed observed. As with the more common nuclei, ^1H and ^{13}C, the phosphorus coupling constants decrease with an increase of bonds between the coupling nuclei. However, it is frequently observed that $^3J_{PX}$ coupling constants are larger than $^2J_{PX}$ coupling constants. In the following we will examine the trends in the coupling constants of phosphorus with the more common nuclei ^1H, ^{13}C, and ^{31}P.

2.2.1 Coupling to Hydrogen

As with other nuclei, $^nJ_{PH}$ coupling constants have a tendency to decrease when the value of n increases, though exceptions are known. Although $^1J_{PH}$ values are frequently in the range of 400–1000 Hz, the coupling constants drop rapidly and are detectable for $^4J_{PH}$ only in special circumstances.

2.2.1.1 $^1J_{PH}$ Coupling Constants

The value for $^1J_{PH}$ increases with decreasing electron density on phosphorus. In the series PH_4^+, PH_3, and PH_2^-, the $^1J_{PH}$ coupling constants are 547 Hz, 189 Hz, and 139 Hz, respectively. It appears that the influence of electron lone pairs on the $^1J_{PH}$ coupling constant becomes smaller with each lone pair. Δ $^1J_{PH}$ for PH_4^+/PH_3 is 358 Hz, whereas for PH_3/PH_2^- Δ, $^1J_{PH}$ is only 50 Hz.

The $^1J_{PH}$ coupling constant shows the same trend upon coordination. When the phosphorus binds to a σ-acceptor, such as a transition metal or a main group Lewis acid, it does so with its lone pair. As a result, the electron density on phosphorus drops sharply and the $^1J_{PH}$ coupling constant increases accordingly. A good example is observed with PH_3, which has a $^1J_{PH}$ coupling constant of 189 Hz as a free ligand, but a significantly larger one in its complexes: 307 Hz in $[Cr(CO)_4(PBu_3)(PH_3)]$, and 366 Hz in Me_3BPH_3, respectively.

Note: *PF$_2$H ($^1J_{PH} = 181.7$ Hz) has a smaller $^1J_{PH}$ coupling constant than PH$_3$ ($^1J_{PH} = 189$ Hz) despite the greater electronegativity of fluorine compared to hydrogen. The reason, of course is π-backbonding (hyperconjugation) observed in P-F, but not in P-H bonds, decreasing the effective electronegativity of fluorine towards phosphorus.*

Similarly, the $^1J_{PH}$ coupling constant increases with an increasing sum of the electronegativities of the substituents on phosphorus. The coupling constant increases in the following order of substituents: H < alkyl < aryl < OR < OPh.

Note: *The $^1J_{PH}$ coupling constant increases with decreasing steric bulk of the substituents on phosphorus.*

This phenomenon is also observed with $^1J_{PC}$ coupling constants and explained there.

2.2.1.2 $^2J_{PH}$ and $^3J_{PH}$ Coupling Constants

In $^2J_{PH}$ and $^3J_{PH}$, coupling, the phosphorus and hydrogen atoms are separated by one (^2J) or two (^3J) atoms of the backbone – typically carbon atoms. This arrangement leads to a Karplus-like dependance of the $^2J_{PH}$ and vicinal $^3J_{PH}$ coupling constants on the dihedral angle Θ.

Note: *The dihedral angle Θ is not the only factor influencing the coupling constant. There exists a separate relationship for each structural class of compound.*

An important contributing factor is the electron density on phosphorus. A dependance on the existence of lone pairs or heteroatoms attached to phosphorus is therefore obvious. The common feature is a minimum of the coupling constant around $\Theta = 90°$ with maxima at $\Theta = 0°$ and $\Theta = 180°$, respectively. Typically, the maximum at $\Theta = 180°$ is larger than at $\Theta = 0°$.

Note: *The H-N-P structural motif usually results in very broad doublets due to the quadrupolar moment of the dominant nitrogen atom ^{14}N ($I = 1$; nuclear abundance $= 99.63\%$).*

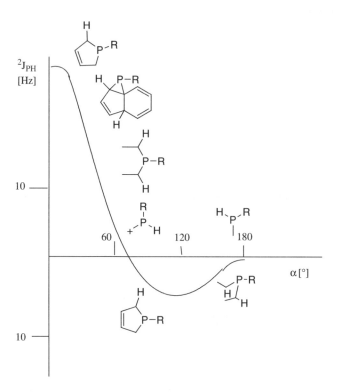

Fig. 2.10 Typical Karplus graph for phosphanes

The relationship between the magnitudes of $^2J_{PH}$ (*gem*) and $^3J_{PH}$ (*cis* and *trans*) coupling constants is similar to that observed in J_{HH} coupling constants. $^3J_{PH}$ (*trans*) is frequently larger than $^3J_{PH}$ (*cis*), but $^2J_{PH}$ (*gem*) can be larger than $^3J_{PH}$ (*trans*). The electron density on phosphorus has a pronounced effect on $^2J_{PH}$ (*gem*) and $^3J_{PH}$ (*cis*). Coordination to a transition metal markedly increases both, whereas electronegative substituents on phosphorus effect only $^3J_{PH}$ (*cis*) in a variable manner. The effect on $^3J_{PH}$ (*trans*) is small, but nonetheless significant.

The coupling constants for $^3J_{PH}$ (*trans*) are typically 10–30 Hz, $^2J_{PH}$ (*gem*) up to 25 Hz, and $^3J_{PH}$ (*cis*) up to 15 Hz. Without electronegative substituents on phosphorus, the values for $^2J_{PH}$ (*gem*) and $^3J_{PH}$ (*cis*) remain well under 10 Hz.

2.2.1.3 $^4J_{PH}$ Coupling Constants

As with $^4J_{HH}$ coupling constants, there is a similar "W" effect for $^4J_{PH}$ coupling meaning that the $^4J_{PH}$ coupling constant reaches a maximum and becomes detectable when H—A_1—A_2—A_3—P (A_i: backbone atoms) adopts a rigid "W" configuration. For a rigid "W" configuration, the H—A_1—A_2—A_3—P sequence in the molecule must be part of a heterocycle or double bond system. Rigid "W" configurations are relatively rare in ^{31}P-NMR spectroscopy.

2.2.2 Coupling to Carbon

The coupling of phosphorus to carbon follows the general trends observed with other nuclei. However, it should be mentioned that the correct assignment of $^nJ_{PC}$ couplings requires a detailed understanding of the relative magnitudes of $^1J_{PC}$, $^2J_{PC}$ and $^3J_{PC}$. The magnitude of $^nJ_{PC}$ is dependent on the oxidation state and the coordination number of phosphorus, the value of n – although it is not a linear correlation – and the stereochemical disposition of coupling nuclei. The electronegativity of substituents on phosphorus also plays an important role.

2.2.2.1 $^nJ_{PC}$ Coupling Constants

In trialkyl phosphanes, the $^1J_{PC}$ and $^3J_{PC}$ coupling constants are usually of similar magnitude, whereas the $^2J_{PC}$ coupling constants are significantly larger. Coupling constants in aryl phosphanes are even more difficult to assign correctly. Table 2.6 shows a few representative examples.

The coupling constant to the *ipso*-carbon of the phenyl ring ($^1J_{PC}$) is frequently smaller than the coupling constant to the *ortho*-carbon ($^2J_{PC}$). However, this is by no means a general trend, as there are numerous examples where $^2J_{PC}$ is smaller than $^1J_{PC}$.

Table 2.6 $^{n}J_{PC}$ coupling constants (P-Ph) of representative aryl phosphines [Hz]

	$(PhCH_2)_2PPh$	$PhCH_2(Ph)_2P$	Ph_3P	Ph_4P^+	$Ph_3P=O$
$^{1}J_{PC}$ [Hz]	20.6	16.2	−12.5	88.4	104.4
$^{2}J_{PC}$ [Hz]	19.8	18.9	19.6	10.9	9.8
$^{3}J_{PC}$ [Hz]	7.0	6.4	6.8	12.8	12.1
$^{4}J_{PC}$ [Hz]	0.6	0.2	0.3	2.9	2.8

Note: *The magnitude of $^{1}J_{PC}$ increases dramatically upon loss of the lone pair on phosphorus.*

Note: *$^{3}J_{PC}$ becomes larger than $^{2}J_{PC}$ upon loss of the lone pair on phosphorus.*

For methyl phosphorus derivatives, the $^{1}J_{PC}$ coupling constant for compounds with phosphorus in the oxidation state +III is generally found to be small or even negative, whereas the corresponding $^{1}J_{PC}$ coupling constant for phosphorus compounds where the phosphorus is in the oxidation state +V is usually substantially larger and positive. This trend is not surprising, as oxidation results in a lower electron density on phosphorus, and thus an increased $^{1}J_{PC}$ coupling constant.

Upon coordination to transition metals, the $^{n}J_{PC}$ coupling constants in trialkyl phosphanes are influenced differently. The $^{1}J_{PC}$ coupling constant increases substantially, the $^{2}J_{PC}$ coupling constant decreases substantially, and the $^{3}J_{PC}$ coupling constant increases slightly. A typical example is shown in Fig. 2.11.

2.2.2.2 Tetracovalent Phosphorus

In ^{31}P-NMR spectroscopy, the term tetracovalent phosphorus is sometimes used with a different meaning. Here, the phosphorus lone pair is counted as a valency, although strictly speaking that is incorrect. However, counting the lone pair, a trialkyl phosphane features a tetracovalent phosphorus atom. That phosphorus atom should be sp^3-hybridized.

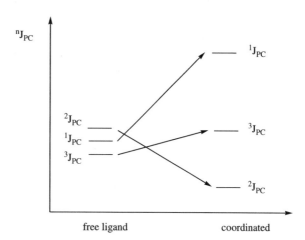

Fig. 2.11 Influence of coordination on $^{n}J_{PC}$

Increase of steric bulk of the substituents on phosphorus causes an increase in $^1J_{PC}$, irrespective of sign. This observation is frequently interpreted as evidence for an increase in s-character in the PC bond accompanied by an increase in the CPC bond angles. The increase in CPC bond angles due to steric crowding on phosphorus is confirmed by experimental evidence. Let us consider PMe_3 and PBu^t_3 as examples. The former has a CPC bond angle of 98.8° and a $^1J_{PC}$ coupling constant of −13.6 Hz, whereas the latter, much bulkier, phosphane has a CPC bond angle of 109.9° and a much smaller $^1J_{PC}$ coupling constant of −33.9 Hz. PMe_3 can thus be seen as having an s-character lone pair, whereas PBu^t_3 is sp^3-hybridized.

The situation can also be explained without MO theory, looking instead towards Gillespie's VSEPR (valence shell electron pair repulsion theory). According to VSEPR, the substituents on phosphorus (including the "stereoactive" lone pair) form a tetrahedron with the lone pair occupying more space than the PC bonds. This results in a CPC bond angle considerably smaller than the tetrahedral angle of 109.5°. Steric bulk on the carbon substituents causes repulsion between the groups, and consequently an increase of the CPC bond angle at the expense of the lone pair. The coupling constant depends now on the C-P-C bond angles without reference to changes in s-character.

Note: *Increase of steric bulk around phosphorus decreases the $^1J_{PC}$ coupling constant. $^1J_{PC}$ coupling constants are frequently negative.*

Note: *In first order NMR-spectra, the phase information on the coupling constant is lost, and thus all coupling constants appear to be positive.*

2.2.2.3 Pentacoordinate Phosphorus

The VSEPR theory tells us that a pentacoordinate phosphorus atom features a trigonal bipyramidal geometry. There are two distinct sites on a trigonal bipyramid, the equatorial and the axial sites. The equatorial sites can be described as utilizing sp^2-orbitals, and the axial sites the remaining p-orbital. That would mean that the axial bonds have no s-character, and thus feature smaller $^1J_{PC}$ coupling constants than the equatorial sites.

Normally, no differences in the $^1J_{PC}$ coupling constants are observed in compounds with pentacoordinated phosphorus since these compounds are subject to a fast exchange of their equatorial and axial sites known as pseudorotation, or Berry rotation (see Fig. 2.12). For different $^1J_{PC}$ coupling constants to be observed, the

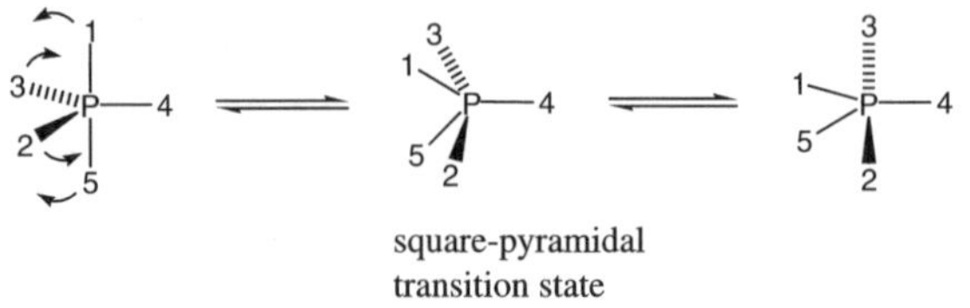

Fig. 2.12 Depiction of the pseudorotation or Berry rotation

square-pyramidal transition state

Berry rotation has to be prevented, usually by performing the experiment at lower temperatures.

Note: *In a trigonal bipyramidal geometry, two different $^1J_{PC}$ coupling constants are only observed at temperatures low enough for Berry rotation not to occur.*

2.2.3 Coupling to Phosphorus

J_{PP} coupling constants in organophosphorus compounds cover a wide range of values. The absolute values have to be treated with caution, since the FT-NMR experiment yields the coupling constant as the difference between two lines and does not give the sign. Many authors report this difference as a positive number without establishing the correct sign for the coupling constant.

From the available information (see Figs. 2.13 and 2.14), we can deduce that steric crowding decreases the $^1J_{PP}$ coupling constant significantly. Whereas P_2Me_4

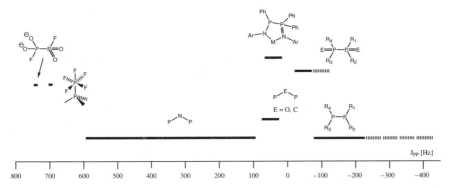

Fig. 2.13 Selected $^1J_{PP}$ and $^2J_{PP}$ coupling constants. The dashed line marks the influence of steric factors

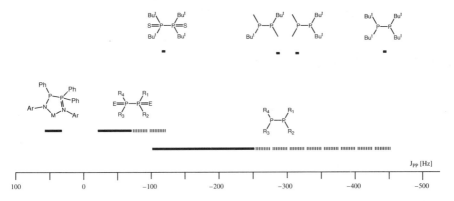

Fig. 2.14 The steric contribution toward $^1J_{PP}$ coupling constants in diphosphanes

has a coupling constant of $^1J_{PP}=-180\,Hz$, this value drops to $^1J_{PP}=-290\,Hz$ for Bu^t-$MePPMeBu^t$, $^1J_{PP}=-318\,Hz$ for $Me_2PPBu^t_2$, and as low as $^1J_{PP}=-451\,Hz$ for $P_2Bu^t_4$. The corresponding doubly oxidized derivatives $R_2(E)PP(E)R_2$ (E=O, S, Se) follow the same trend, but they can be found to the left (larger coupling constants) from the non-oxidized analogues. $P_2Me_4S_2$ has a coupling constant of $^1J_{PP}=-19\,Hz$, and $\{P(S)MeBu^t\}_2$ a coupling constant of $^1J_{PP}=-109\,Hz$. Finally, the monoxidized diphosphanes $R_2(E)PPR_2$ (E=O, S, Se) have similar $^1J_{PP}$ coupling constants to the diphosphanes, indicating that one PALP is sufficient for the magnitude of the $^1J_{PP}$ coupling constant.

Note: *Steric crowding results in a decrease of the $^1J_{PP}$ coupling constant in diphosphanes.*

Note: *Lower electron density results in an increase of $^1J_{PP}$ constants in diphosphanes.*

Phosphorus phosphorus coupling through metal atoms shows a few interesting aspects. There are geometrical isomers in octahedral and square planar metal complexes, namely *cis*- and *trans*-coordination. *Trans*-complexes usually show a larger $^2J_{PP}$ coupling constant than *cis*-complexes. The explanation seems simple. From the Karplus diagrams, we know that a 180° dihedral angle results in a larger coupling constant than a 90° dihedral angle. Although *cis/trans* isomers do not possess a dihedral angle, it is tempting to assume the same angular dependance. Another explanation cites coupling through the same metal d-orbital for *trans*-complexes. Unfortunately, the explanation is not so simple, but we can content ourselves with the observation that $^2J_{PP}(trans)$ is generally larger than $^2J_{PP}(cis)$ and $^2J_{PP}(trans)$ is significantly larger for 4d and 5d metals than for 3d metals.

There is also a significant difference in J_{PP} coupling between the metal complexes of monodentate and chelate phosphanes. Monodentate phosphanes can only couple through the metal, whereas chelate phosphanes can couple through the metal, and additionally through the backbone.

Note: *coupling through the backbone increases (same sign) or diminishes (opposite sign) the phosphorus phosphorus coupling constant relative to analogous monodentate phosphanes, depending on whether the two possible coupling routes are in phase or not.*

2.2.4 Coupling to Metals

Most metals have quadrupolar nuclei. For these, J_{PM} coupling constants are measured by CP/MAS ^{31}P-NMR spectroscopy in the solid state. This technique is not within the scope of this book. We will limit ourselves to the dipolar nuclei ^{103}Rh and ^{195}Pt.

Note: *the $^1J_{PM}$ coupling constant increases with increasing s-character of the M-P bond.*

Note: *in tbp complexes $^1J_{PM}$ is larger for equatorial than for axial ligands.*

Note: *the $^1J_{PM}$ coupling constant decreases as the oxidation state for platinum complexes increases.*

Note: *phosphite ligands cause a 50–100% larger $^1J_{PM}$ coupling constant compared to analogous phosphane ligands.*

Note: *$^1J_{PtP}$ correlates strongly with the platinum phosphorus bond length.*

Bibliography

Albrand J P, Gagnaire D, Robert J B, J Chem Soc, Chem Comm (1968) 1469.

Berry R S, J Chem Phys 32 (1960) 933.

Gillespie R J, J Chem Educ 40 (1963) 295.

Gillespie R J, J Chem Educ 47 (1970) 18.

Kempgens P, Elbayed K, Raya J, Granger P, Rose J, Braunstein P, Inorg Chem 45 (2006) 3378.

Klapötke T M in Riedel E [ed], Moderne Anorganische Chemie, Walter de Gruyter, Berlin, 3. Auflage 2007.

Nelson J H, Nuclear Magnetic Resonance Spectroscopy, Prentice Hall, Upper Saddle River, NJ (USA) 2003.

Quin L D, Verkade J G, Phosphorus-31 NMR Spectral Properties in Compound Characterization and Structural Analysis, VCH, Weinheim, 1994.

Reed A E, von Ragué Schleyer P, J Am Chem Soc 112 (1990) 1434.

Richet T, Elbayed K, Raya J, Braunstein P, Rose J, Magn Res Chem 34 (1996) 689.

Tebby J [ed], Handbook of Phosphorus-31 Nuclear Magnetic Resonance Data, CRC Press, Boca Raton, USA, 1990.

Verkade J G, Quin L D [eds], ^{31}P-NMR Spectroscopy in Stereochemical Analysis, VCH, Weinheim, 1987.

Von Ragué Schleyer P, Kos A J, Tetrahedron 39 (1983)1141.

Chapter 3
Oxidation State from P(–III) over P(0) to P(+V)

As chemists, we are used to examining the electronic structure of compounds in terms of the oxidation states of the constituting atoms. The concept of oxidation states works very well in Synthetic Chemistry and explains the reactivity of functional groups as well as the mechanisms of reactions.

But does it reflect the phosphorus chemical shift accurately? To answer this question, we might want to revisit the definition of the oxidation state and how it is assigned.

The starting point in the assignment of the oxidation state of a specific atom in a given compound is the number of valence electrons that this atom has as an element. For phosphorus, this number is five. Next, we look at the electronegativity of this atom, 2.19 for phosphorus, and then at the electronegativities of the atoms it bonds to. Normally, in phosphorus chemistry, these are hydrogen (2.2), carbon (2.55), nitrogen (3.04), oxygen (3.44), sulfur (2.58), fluorine (3.98), and chlorine (3.16). We also encounter a variety of metals, all of which essentially have an electronegativity smaller than that of phosphorus.

Note: *The bonding electrons are assigned entirely to the more electronegative atom.*

This means that in PPh_3, one of the most popular phosphanes, three of the five electrons of phosphorus in its native state have been transferred to the carbon substituents, leaving the phosphorus atom with two electrons and the oxidation state +III ($5-2=+III$).

Note: *Donor bonds do not change the oxidation state.*

Donor bonds are usually formed between an electronegative and electron rich non-metal atom possessing a lone pair, and a metal or semi-metal atom as a Lewis acid with an empty orbital. The electrons are still deemed to belong to the Lewis base, in full accord with the oxidation state concept. However, it is easy to understand that in reality, transfer of electron density from the Lewis base to the Lewis acid has occurred.

Note: *Oxidation states do not reflect the true electron density on phosphorus.*

If it is true that oxidation states oversimplify the electron transfer trends in a given molecule, and thus the electron density on phosphorus, then it follows that the oxidation state concept cannot be used to explain or predict phosphorus chemical shifts.

O. Kühl, *Phosphorus-31 NMR Spectroscopy*,
© Springer-Verlag Berlin Heidelberg 2008

Note: *^{31}P-NMR chemical shifts are independent of formal oxidation numbers. Rather, they depend on the actual electron density on phosphorus, the bond angles, and π-bonding.*

We now have to prove our assumption by correlating phosphorus chemical shifts with the oxidation state of phosphorus in a range of molecules. If the oxidation state correlates with the chemical shift, then we will find well-defined blocks of molecules that experience a downfield shift with increasing oxidation state. If our assumption is true, however, then we will find that for each oxidation state, the constituting molecules will cover a broad range of chemical shift values, and that these blocks have large overlap areas.

In evaluating the dependance of the phosphorus chemical shift on the oxidation state of phosphorus in the molecule, we will make a few assumptions regarding electronegativities. There is an uncertainty concerning the relative electronegativities of phosphorus and hydrogen. The scale of Alred-Rochow has hydrogen more electronegative (by 0.01 units) than phosphorus, whereas Pauling has them at equal values. We will regard phosphorus as more electronegative than hydrogen on the grounds that phosphanes react with strong bases under proton abstraction. Equally, we regard all metals, including the "platinum metals," as less electronegative than phosphorus.

Now, we turn to a detailed evaluation of the phosphorus resonances where phosphorus is in the oxidation state –*I* or +*I*. A graphical description is given in Fig. 3.1.

We find that the oxidation state +*I*, with a chemical shift range of $\Delta\delta = 721$ ppm (from $\delta_p = -231$ ppm to $\delta_p = 490$ ppm), is completely embedded in the chemical shift range for the oxidation state –*I* of $\Delta\delta = 1714$ ppm (from $\delta_p = -352$ ppm to $\delta_p = 1362$ ppm). The change of oxidation state apparently makes no difference to

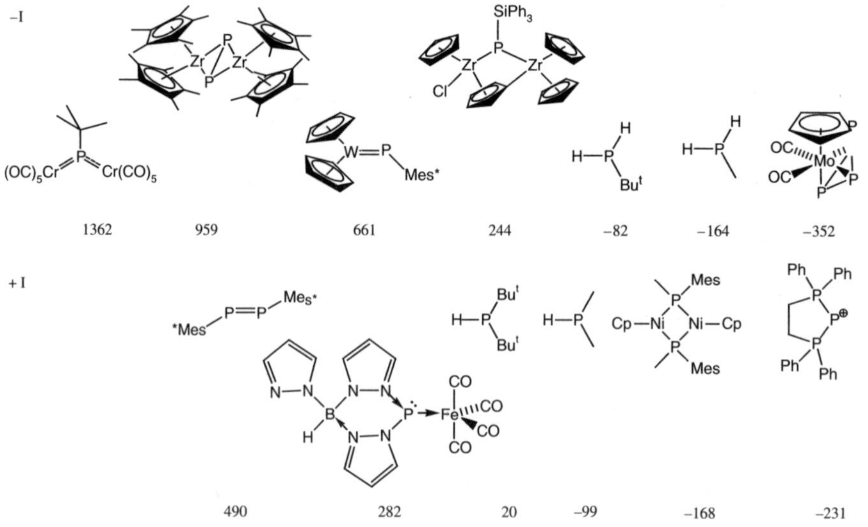

Fig. 3.1 The ^{31}P chemical shift ranges for P(*I*) and P(–*I*)

the chemical shift. If we look at Fig. 3.2, where individual chemical shift ranges are depicted for all eight oxidation states of phosphorus (−III to +V), we instantly realize that the oxidation state obviously does not contribute to the chemical shift value in ^{31}P-NMR spectroscopy.

A Cautionary Note: *^{31}P-NMR spectroscopy books written by analytical chemists frequently list phosphonium salts as compounds with phosphorus in the oxidation state +IV.* **That is incorrect**. *Phosphonium salts are synthesised according to the following reaction:*

$$PPh_3 + Ph\!-\!I \rightarrow PPh_4^+ + I^-$$

The phosphorus lone pair attacks the C-I bond, and subsequently an iodide anion leaves the transition state. The resulting phosphonium cation has four P-C bonds and a positive charge on phosphorus, giving phosphorus the oxidation state +V and not +IV.

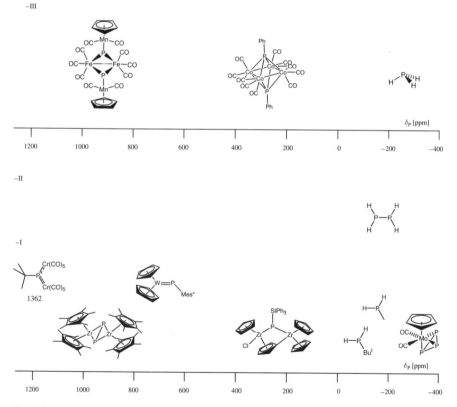

Fig. 3.2 The "dependance" of the ^{31}P chemical shift on the phosphorus oxidation state

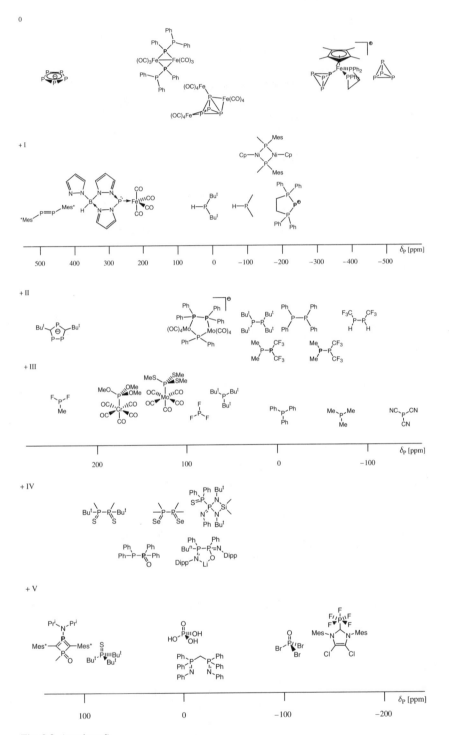

Fig. 3.2 (continued)

Phosphorus does realize the oxidation state +IV; in fact, there are examples for all nine possible oxidation states from −III to +V. The phosphorus (IV) compounds are known as diphosphanedioxides $R_2(O)PP(O)R_2$ and diphosphanemonoxide $R_2PP(O)R_2$. The oxygen atom(s) can be replaced by another atom such as sulfur, selenium, or nitrogen.

Bibliography

Baudler M, Angew Chem 94 (1982) 520.
Baudler M, Angew Chem 99 (1987) 429.
Baudler M, Glinka K, Chem Rev 94 (1994) 1273.
Quin L D, A Guide to Organophosphorus Chemistry, Wiley-VCH, Weinheim, 2000.

Chapter 4
λ^5-Phosphanes

The λ^5-phosphanes, with phosphorus in the oxidation state +V, have chemical shift values around $\delta_p=0$ ppm. This follows from the definition of phosphorus chemical shifts – the standard for $\delta_p=0$ ppm is orthophosphoric acid, H_3PO_4, a P (+V) compound.

We expect to see P (+V) compounds with positive (downfield) and negative (upfield) chemical shifts depending on the substituents on phosphorus. However, what decides the direction and the magnitude of the additional chemical shift? From the standpoint of electronegativity, substituting the OH-groups in orthophosphoric acid, H_3PO_4 by alkyl or aryl groups would decrease the sum of electronegativity of the substituents, and thus should increase the electron density on phosphorus and result in an upfield shift. The data in Table 4.1 (left column), however, consistently show downfield shifts for this compound class. If we exchange the hydrogen atoms in H_3PO_4 with alkyl or aryl groups, then we see upfield or downfield shifts, after expecting downfield shifts on the grounds of electronegativity considerations.

Putting the data from Table 4.1 in order of decreasing chemical shift values (upfield shift), we make the following observations:

For O=PR$_3$: δ_p: Bu^t>Et>Bu^n>Bz>Me>Ph or for alkyl Bu^t>Et>Bu^n>Me

For O=P(OR)$_3$: δ_p: Bu^n>Me>Bz>Et>Bu^t>Ph or for alkyl Bu^n>Me>Et>Bu^t

or in reference to the two phosphorus compound classes depicted in Table 4.1:

Note: *If the alkyl or aryl substituent is bonded to oxygen, it has the opposite effect as if it were bonded to phosphorus directly. On oxygen, it acts upfield, on phosphorus, downfield.*

That leads us to the following explanation: Carbon is more electronegative than phosphorus, but less so than oxygen. In consequence, carbon substitution on oxygen increases the electron density at oxygen and results in less electron density withdrawal from phosphorus in the case of O=P(OR)$_3$. That translates into an upfield shift with increased carbon substitution. For O=PR$_3$, the situation is different. Since carbon is more electronegative than phosphorus, increased carbon substitution on phosphorus results in decreased electron density on phosphorus, and thus a downfield shift.

O. Kühl, *Phosphorus-31 NMR Spectroscopy*,
© Springer-Verlag Berlin Heidelberg 2008

Table 4.1 ^{31}P chemical shift values for selected phosphane oxides and phosphates

Molecule	δ_P [ppm]	Molecule	δ_P [ppm]
O=PMe$_3$	36.2	O=P(OMe)$_3$	2.1
O=PEt$_3$	48.3	O=P(OEt)$_3$	−1.0
O=PBu$^n_3$	43.2	O=P(OBu^n)$_3$	14.0
O=PBu$^t_3$	66.5	O=P(OBu^t)$_3$	−13.3
O=PBz$_3$	40.7	O=P(OBz)$_3$	0.9
O=PPh$_3$	28	O=P(OPh)$_3$	−17.3

The same behavior as in Table 4.1 is observed in Fig. 4.1: although oxygen is more electronegative than carbon, O=PC$_i$Hal$_{3-i}$ is shifted downfield from O=P(OR)$_i$Hal$_{3-i}$. The reason is obviously the same: carbon has a +I effect on oxygen, but a −I effect on phosphorus. However, let us examine the situation within the two compound classes more closely. Substituting carbon for halogen in O=PC$_i$Hal$_{3-i}$ results in a downfield shift when going from O=PC$_3$ to O=PC$_2$Hal, but in upfield shifts for every consecutive substitution.

The halogen is usually more electronegative than the carbon it replaces, resulting in a downfield shift. The halogen is also capable of a π-donor interaction (hyperconjugation) toward phosphorus, resulting in an upfield shift. In the first substitution step, from O=PC$_3$ to O=PC$_2$Hal, the downfield contribution dominates the upfield one, but subsequently, it is the upfield shift that prevails.

In Fig. 4.2, we see another interesting phenomenon. We would expect that O=PR$_3$ would resonate downfield from S=PR$_3$, since oxygen is more electronegative than sulfur. In reality, it is the other way around.

Note: *Phosphane oxides resonate upfield from the respective thio compounds.*

In Fig. 4.3, we see that the effect extends to the iminophosphoranes R$_3$P=NR. Apparently, the second row elements nitrogen and oxygen resonate upfield from the

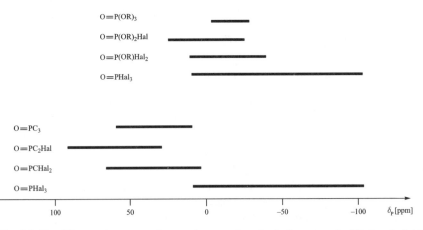

Fig. 4.1 The difference between carbon- and oxygen-based substituents on the ^{31}P chemical shift of selected phosphorus compounds

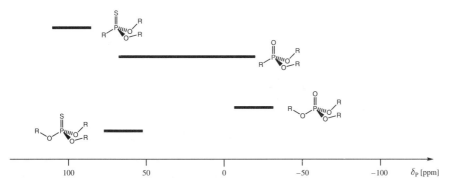

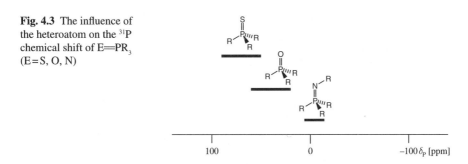

Fig. 4.2 The influence of the chalcogen on the ^{31}P chemical shift of some selected E=PR$_3$ (E=S, O) systems

Fig. 4.3 The influence of the heteroatom on the ^{31}P chemical shift of E=PR$_3$ (E=S, O, N)

third row element sulfur, despite the trend of electronegativities. An explanation can be attempted using the concept of negative hyperconjugation with a VB representation of the bonding as P$^+$–E$^-$ (E=N, O, S).

Note: *The trend is S=PR$_3$ downfield from O=PR$_3$ downfield from RN=PR$_3$.*

As we go from five coordinate (PR$_5$) to six coordinate (PR$_6^-$) P (V) compounds, we observe a significant upfield shift. This is easy to understand. The PR$_5$ moiety is a Lewis acid, and behaves as a σ-acceptor towards the Lewis acidic anion. The electron density on phosphorus increases, and we see an upfield shift in the ^{31}P-NMR spectrum.

This upfield chemical shift is in no way restricted to anionic ligands, as can be seen by the strong σ-donor dimesityl-imidazolydene, an Arduengo carbene (see Fig. 4.5).

That leaves us with the following question: why O=PF$_3$ has a phosphorus chemical shift that is so very far downfield from PF$_5$. Both fluorine and oxygen have similar electronegativities, with fluorine actually being more electronegative than oxygen. Furthermore, both oxygen and fluorine are capable of π-donor interactions toward phosphorus. Could it be that the tetrahedral geometry of O=PF$_3$ is responsible for this downfield shift compared to the trigonal bipyramidal geometry of PF$_5$?

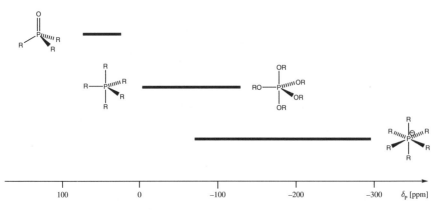

Fig. 4.4 The influence of the coordination geometry on the ^{31}P chemical shift

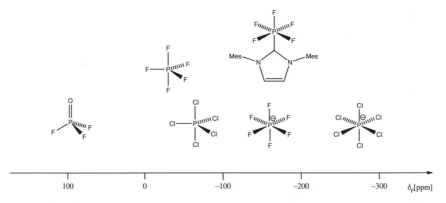

Fig. 4.5 The ^{31}P chemical shifts of some selected P(V) compounds

The trigonal bipyramidal geometry has some interesting electronic features. There are two distinct positions in a trigonal bipyramid, or tbp: the axial and the equatorial positions (see Fig. 4.6). The equatorial position is more electron donating than the axial position. As a consequence, substituents capable of π-donor interactions are often found in the equatorial position. Seen from the phosphorus atom's point of view, adopting a trigonal bipyramidal geometry with π-donating substituents in the equatorial position results in increased electron density on phosphorus. Since the tpb geometry is subject to constant Berry rotation, substituent scrambling is often observed resulting in an average occupancy of the axial and equatorial positions for each substituent.

Note: *In a trigonal bipyramidal geometry, the central atom gains more electron density from its equatorial substituents than the central atom of a tetrahedral geometry.*

The upfield shift of PF$_5$ compared to O=PF$_3$ is at least partially due to the trigonal bipyramidal geometry compared to the tetrahedral geometry.

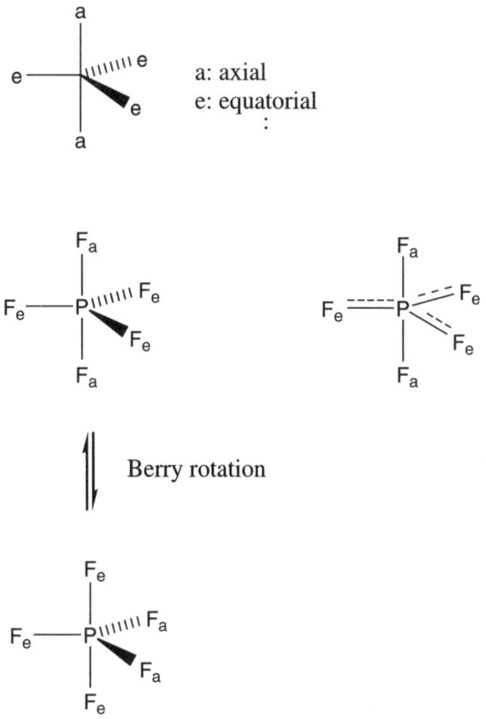

Fig. 4.6 The two distinct coordination sites in a trigonal bipyramide and their rapid exchange through Berry rotation

a: axial
e: equatorial

Berry rotation

The second difference between $O{=}PF_3$ and PF_5 is the existence of a formal P=O double bond (really a P^+—O^- bond subject to hyperconjugation) compared to two P—F single bonds. We will find out in Chap. 5 that a phosphorus element double bond results in a significant downfield shift of the phosphorus resonance. That should not surprise us. We know that a π-donor interaction from a subsituent results in a significant upfield shift. By the same argument, a double bond to a more electronegative element should result in a significant downfield shift, as observed.

Note: *Double bonds toward more electronegative elements result in significant downfield shifts of the phosphorus resonance.*

Bibliography

Demange M, Boubekeur L, Auffrant A, Mezailles N, Ricard L, Le Goff X, Le Floch P, New J Chem 30 (2006) 1745.

Quin L D, Williams A J, Practical Interpretation of P-31 NMR Spectra and Computer-Assisted Structure Verification, Advanced Chemistry Development, Toronto, 2004.

Chapter 5
λ^3-Phosphanes

5.1 Phosphorus Element Single Bonds

The great importance of phosphanes as ligands in organometallic chemistry prompted chemists to search for methods to predict the ^{31}P-NMR chemical shifts of these compounds not only qualitatively, but quantitatively. To this end, the following set of empirically derived equations was developed.

In these equations, σ^c denotes a shift constant characteristic for a certain substituent. The values are given in Table 5.2. The parameters a and b stand for the number of allyl, benzyl, or cyclohexyl groups (a), and phenyl groups (b), bonded to the phosphorus atom of a phosphonium salt.

The set of equations given in Table 5.1 are the only ones developed for the prediction of ^{31}P-NMR chemical shifts of phosphanes. There is another set of equations for transition metal complexes. This lack of predictive tools reflects the multitude of factors influencing and determining the ^{31}P-NMR chemical shift values of phosphorus compounds. Three main factors have been identified in the bond angle, the electronegativity of the substituents, and the π-bonding character of these substituents. It is the electronegativity argument in the absence of the other two factors that gives rise to the equations in Table 5.1.

Note: *The π-bonding character of the substituents frequently determines the chemical shift value of the phosphane.*

We note that the equations in Table 5.1 do not cover phosphites, one of the most important classes of phosphanes. We also note that phosphites contain alkoxy and/or aryloxy substituents that cause +M effects as substituents on phenyl rings. In other words, they are prone to π-bonding.

Figure 5.1 shows a series of phosphites and thiophosphites. We immediately observe that the trend in downfield shifts for the alkoxy groups is somewhat reverse to that of alkyl ligands. Whereas alkyl substituents on phosphorus cause downfield shifts in the order methyl < ethyl < propyl < butyl, the corresponding alkoxy groups have the order ethyl < *tert*-butyl < *n*-butyl < methyl.

A main factor for this change in the order is the fact that in phosphanes, the alkyl group operates on the phosphorus atom (–I effect, since C is more electronegative

Table 5.1 Equations for the calculation of ^{31}P chemical shifts of certain phosphanes and phosphonium salts

Primary phosphanes	RPH$_2$:	$\delta_p = -163.5 + 2.5\,\sigma^c$
Secondary phosphane	R$_2$PH:	$\delta_p = -99 + 1.5\sum\limits_{n=1}^{2}\sigma^c_n$
Tertiary phosphane	R$_3$P:	$\delta_p = -62 + \sum\limits_{n=1}^{3}\sigma^c_n$
Quaternary phosphonium salt	R$_4$P$^+$:	$\delta_p = 21.5 + 0.26\sum\limits_{n=1}^{2}\sigma^c_n - 3.22\,a - 5.5\,b$

Table 5.2 Shift constants used for the equations in Table 5.1

Substituent	σ^c [ppm]	Substituent	σ^c [ppm]	Substituent	σ^c [ppm]
methyl	0	*neo*-pentyl	3	CH$_2$C≡CH	13
ethyl	14	cyclopentyl	21	C≡CH	–10
n-propyl	10	cyclohexyl	23	C≡CCH$_3$	–6
i-propyl	27	benzyl	17	(CH$_2$)$_2$CN	13
i-butyl	6	phenyl	18	CN	–24.5
sec-butyl	24	allyl	9	NEt$_2$	–1.0
tert-butyl	42	vinyl	15	dibenzophosphole	37
tert-amyl	42	*cis*-CH$_2$(CH)$_2$CH$_3$	–6		
n-C$_n$H$_{2n+1}$; n>3	10	*trans*-CH$_2$(CH)$_2$CH$_3$	9		

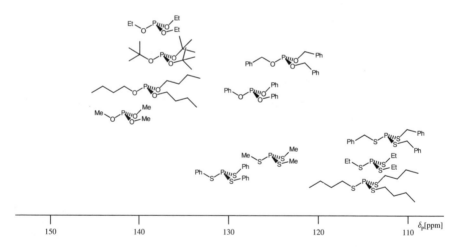

Fig. 5.1 ^{31}P chemical shifts of a series of phosphites and thiophosphites

than P), and in phosphites it operates on the oxygen atom (+I effect, since C is less electronegative than O). An increased electron density on oxygen results in a reduced electron withdrawing effect on phosphorus.

Note: *Carbon substituents in phosphanes cause downfield shifts, whereas on oxygen in phosphites, they cause relative upfield shifts.*

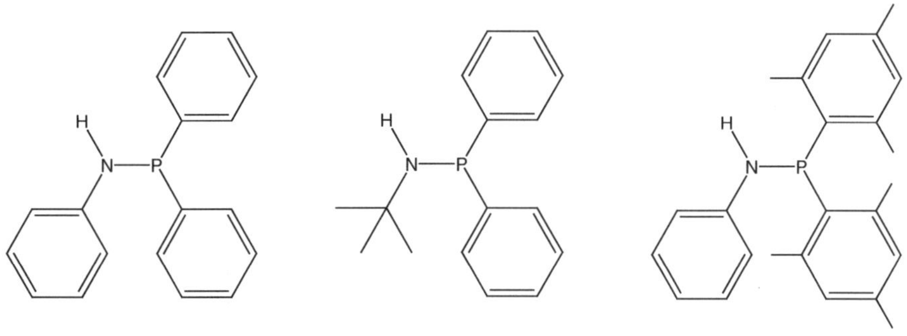

$\delta_P = 28.6$ ppm   22.5 ppm   19.4 ppm

Fig. 5.2 A series of phosphino amines

The oxygen atom is capable of π-bonding interactions, and these tend to operate independently from the electronegativity effect of the substituents. The result is the somewhat erratic order of chemical shift values displayed in Fig. 5.1. The situation for the thiophosphites is similar.

Note: *The oxygen and sulfur atoms of phosphites and thiophosphites can engage in π-donor interactions causing upfield chemical shifts.*

If we substitute a carbon substituent of a tertiary phosphane by an amino group, we would expect a downfield shift simply because nitrogen is more electronegative than carbon. In Fig. 5.2, we see phosphorus chemical shifts of $\delta_P = 19.4$ ppm to $\delta_P = 28.6$ ppm downfield of most tertiary phosphanes. We note that all three phosphino amines depicted in Fig. 5.2 still have an NH functionality. We suspect that deprotonation will result in a further upfield shift.

Note: *Phosphino amines have phosphorus chemical shifts downfield from the respective phosphanes.*

What we have said in Chap. 4 about substitution on phosphanoxides should also be applicable here. Carbon substitution on nitrogen operates on nitrogen first, meaning that a tert-butyl group is responsible for an upfield shift compared to a phenyl group, as is indeed observed. Deprotonation of the NH group creates an amide with a negative charge, and thus increased electron density on the nitrogen atom. This results in an upfield shift of the phosphorus resonance.

Note: *Lithiation of a phosphino amine can cause a downfield chemical shift due to coordination of the nitrogen atoms to two lithium atoms each.*

Deprotonation of NHPhPPh$_2$ using BuLi generates the expected phosphino amide [Li(thf)NPhPPh$_2$]$_2$ (see Fig. 5.3). The ^{31}P-NMR chemical shifts are in reverse order to our initial expectation. The phosphino amide [Li(thf)NPhPPh$_2$]$_2$ resonates downfield, not upfield, from the corresponding phosphino amine NHPhPPh$_2$. This might be surprising, but is readily explained upon examining the structure of [Li(thf)NPhPPh$_2$]$_2$. The amide is a dimer whereby each lithium atom bridges two

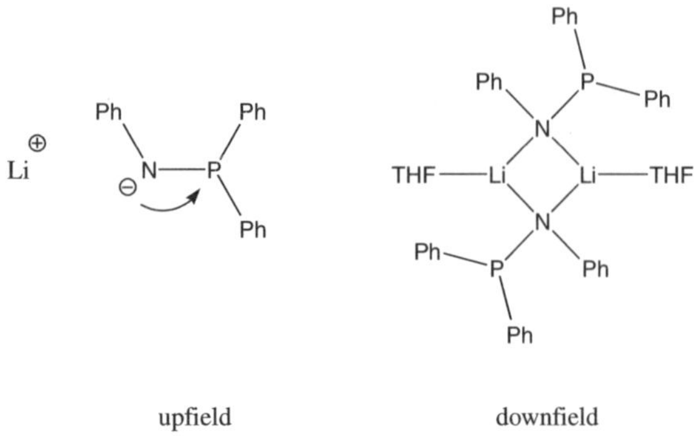

Fig. 5.3 Possible structures
of a lithium phosphino amide
and the expected effect on the
phosphorus chemical shift
for each

upfield   downfield

nitrogen centres. The nitrogen atom engages both lone pairs in donor bonds to lithium, compared to a solitary covalent bond to hydrogen in the amine. The nitrogen atom in the amide therefore has a lower effective electron density than in the amine. The result is the observed downfield shift.

Not all phosphino amides resonate downfield from their respective phosphino amines. In Fig. 5.4, we see an example for an upfield shift of $\Delta\delta=-12$ ppm upon deprotonation. The upfield shift is very modest due to intermolecular hydrogen bonding (explained below). In the heterocyclic structure of the amide, the additional electron density due to deprotonation stays on the nitrogen atom. The potassium atom is coordinated by the oxygen and sulfur atoms.

We remember from organic chemistry that amino groups can be protected by acylation. Transformation of an amine to the corresponding carboxylic acid amide reduces the electron density on nitrogen, and thus the reactivity of the nitrogen atom. In terms of ^{31}P-NMR spectroscopy, the reduced electron density on nitrogen should result in a downfield shift upon acylation of the amino group in a phosphino amine like NHPhPPh$_2$. This is indeed the case (see Fig. 5.5). The phosphino amine NHPhPPh$_2$ has a phosphorus chemical shift of $\delta_p=22.5$ ppm, some $\Delta\delta=30$ ppm upfield from the acylated compound PPh$_2$MeNC(O)Me, a derivative of acetic acid, and (NMePPh$_2$)$_2$CO, a derivative of urea.

Note: *N-acylation of phosphino amines causes a downfield shift in the phosphorus resonance.*

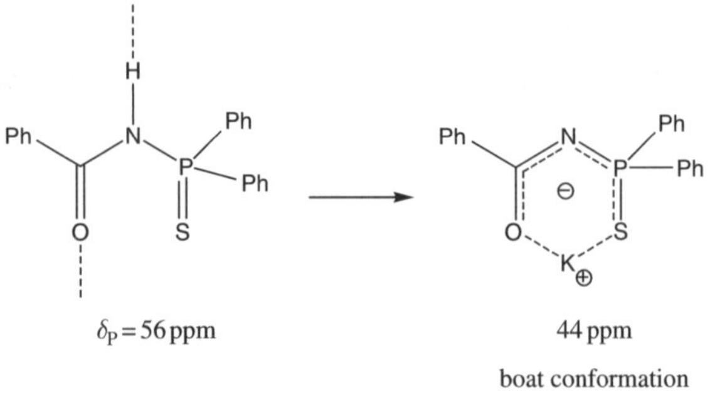

Fig. 5.4 The effect of $\delta_P = 56$ ppm   44 ppm
deprotonation in the absence
of coordination to nitrogen   boat conformation

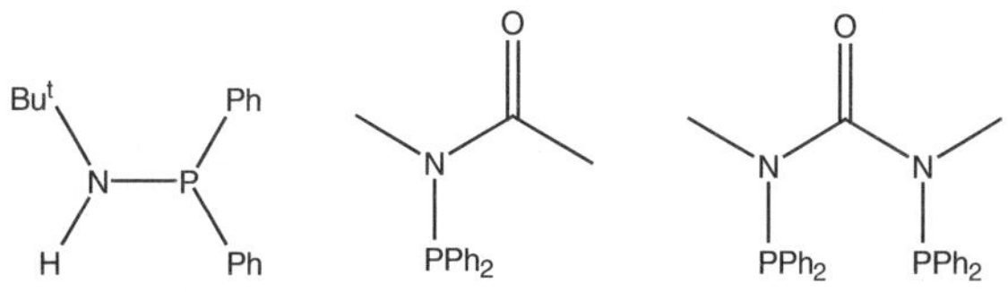

Fig. 5.5 Comparison of phosphorus chemical shifts between phosphino amines, phosphino amides, and phosphino ureas

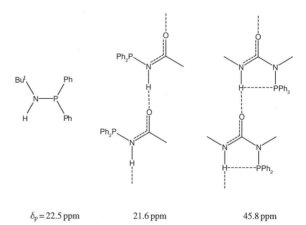

$\delta_P = 22.5$ ppm   55.1 ppm   54.6 ppm

Note: *Reduction of electron density on an atom that is in direct communication with the phosphorus atom usually results in a downfield shift of the phosphorus resonance.*

However, acylation does not always result in a downfield shift of the phosphorus resonance, as we can see in Fig. 5.6. Direct acylation of NH_2PPh_2 yields $PPh_2NHC(O)Me$. The latter has a ^{31}P-NMR chemical shift of $\delta_P = 21.6$ ppm, and thus $\Delta\delta = -0.9$ ppm upfield of the phosphino amine $NHPhPPh_2$. For some reason, the N-acylated phosphino amine $PPh_2NHC(O)Me$, $\delta_P = 21.6$ ppm, has a phosphorus chemical shift $\Delta\delta = -33.5$ ppm upfield from the phosphorus resonance of $PPh_2NMeC(O)Me$, $\delta_P = 55.1$ ppm.

Closer inspection of the two structures reveals a C^1_1 (4) hydrogen bonding pattern. The molecule is part of a $[H—N—C{=}O]^1_\infty$ chain, whereby the molecules are held together by N—H—O=C hydrogen bonds. The hydrogen bonds cause a redistribution of electron density along the H—N—C=O sequence toward an H—N=C—O sequence, as evidenced by a shortened N—C bond. The phosphino group is effectively decoupled from the electronic system of the H—N—C=O chain and the phosphorus chemical shift is that of a non-acylated phosphino amine. The observed chemical shift differences indicate that hydrogen bonding persists in solution.

Note: *Intermolecular hydrogen bonding through the acyl group mitigates the electron withdrawing effect of N-acylation. The phosphorus chemical shift is found upfield from its expected value.*

Fig. 5.6 The influence of hydrogen bonding on the phosphorus chemical shifts

$\delta_P = 22.5$ ppm   21.6 ppm   45.8 ppm

The monophosphino urea compound to the right of Fig. 5.6 has a phosphorus chemical shift intermediate between the bisphosphino urea compound ($NMePPh_2)_2CO$ and the phosphino amine $NHPhPPh_2$ at $\delta_p = 45.8$ ppm. The chemical shift is closer to the compound without hydrogen bonding, although $(NHPPh_2)CO(NMePPh_2)$ possesses a hydrogen donor and two hydrogen acceptor atoms (O and P). Here, the phosphino group acts as a weak intramolecular hydrogen acceptor to the amino group, causing a downfield shift. Since the intermolecular hydrogen bond toward the carbonyl oxygen is still there, the electronic pulling effect is somewhat mitigated, and the downfield shift does not reach its full potential.

We know that phosphane sulfides are generally found upfield from the respective oxides. It interests us whether this is also the case for the N-acylated phosphino amines. In Fig. 5.7, we see a comparison between phosphino urea and phosphino thiourea compounds. It is easy to see that the thiourea derivatives have phosphorus chemical shifts that are slightly downfield from their phosphino urea analogues. The downfield shift is in the range of $\Delta\delta = 5.7$ ppm to $\Delta\delta = 8.9$ ppm. We note that the monophosphino urea/thiourea couple in the middle of Fig. 5.7 is a prominent exception. Here, the thiourea derivative resonates $\Delta\delta = -0.8$ ppm upfield from the urea one. It is evident that the two opposing trends: upfield shift due to intermolecular hydrogen bonding, and downfield shift due to intramolecular hydrogen bonding, are different in the bisphosphino urea and bisphosphino thiourea series.

Note: *In the urea/thiourea system, phosphoramidinato groups can withdraw electron density from the phosphino substituted side. As a result, the phosphoramidinato signal is relatively upfield shifted, and the phosphino signal is relatively downfield shifted.*

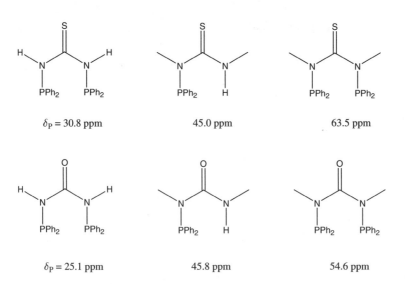

Fig. 5.7 A series of phosphino ureas and phosphino thioureas

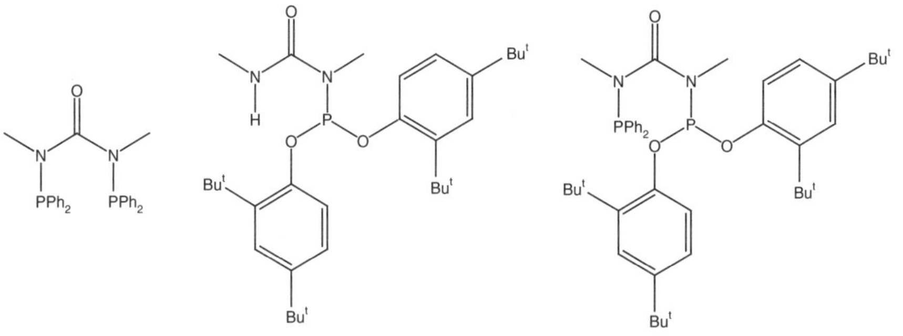

$\delta_p = 54.6\,\text{ppm}$ 128.8 ppm   57.6 ppm   124.6 ppm

Fig. 5.8 The effect of P-aryl and P-aryloxy substituents in phosphino ureas on the phosphorus chemical shift

When we introduce phosphoramidinato groups instead of the phosphino groups in the urea or thiourea compounds, we see a significant downfield shift in the phosphorus resonance. That can be expected, since the chemical shifts for phosphites can be found significantly downfield from the signals for phosphanes. The urea/thiourea system simply mirrors that trend. The monophosphorylated urea species in the middle of Fig. 5.8 shows a phosphorus chemical shift of $\delta_p = 128.8\,\text{ppm}$, whereas the bisphosphorylated compound to the right shows two signals at $\delta_p = 57.6\,\text{ppm}$ for the phosphino group and $\delta_p = 124.6\,\text{ppm}$ for the phosphoramidinato group. This is both significant and surprising. It is surprising since there is obviously no appreciable effect of hydrogen bonding. If anything, one might suspect an intramolecular hydrogen bond to the phosphoramidinato oxygen atom resulting in a small downfield shift of the phosphorus resonance.

It is significant that this upfield shift of $\Delta\delta = -4.2\,\text{ppm}$ for the phosphoramidinato group is accompanied by a downfield shift of $\Delta\delta = 3.0\,\text{ppm}$ for the phosphino group. Unfortunately, the comparison is somewhat reduced in its worth, since the upfield shift of the phosphoramidinato group is measured against a monophosphoramidinato urea. Here, it is fortunate that there is no interference from hydrogen bonding.

The thiourea resonances are once again downfield from their urea counterparts. In Fig. 5.9, we have a complete series of bisphosphino, phosphino phosphoramidinato, and bisphosphoramidinato thiourea derivatives complete with matching substituents on nitrogen and phosphorus. Once again, the mixed substituted compound displays a downfield shift for the phosphino group, $\Delta\delta = 4.9\,\text{ppm}$ compared to the bisphosphino derivative, and a corresponding upfield shift of $\Delta\delta = -2.1\,\text{ppm}$ compared to the bisphosphoramidinato derivative.

The explanation is again rather simple. The phosphino side is *per se* more electron rich than the phosphoramidinato side. Thus, the phosphoramidinato group can withdraw electron density from the phosphino side, causing the observed upfield/downfield shifts.

Fig. 5.9 The effect of
P-methoxy substituents in
phosphino thioureas on the
phosphorus chemical shifts

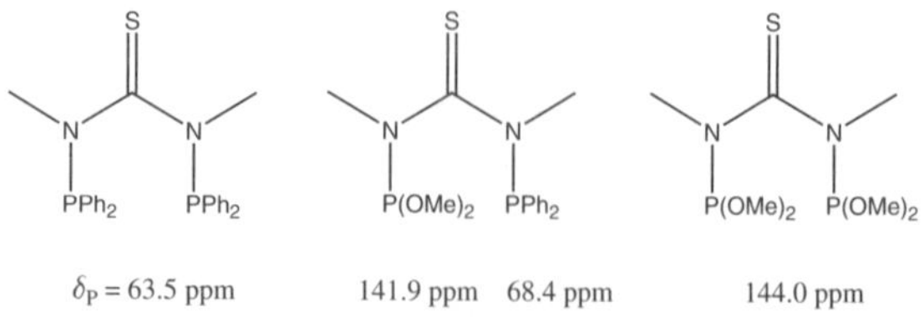

$\delta_P = 63.5$ ppm   141.9 ppm   68.4 ppm   144.0 ppm

Box Story: Electronic and Steric Properties of Phosphanes and Phosphites

The electronic and steric properties of phosphanes are of great interest, since they determine the properties of many catalysts utilizing phosphanes as ligands. Many methods have been devised to evaluate, measure, and calculate these properties, both experimentally and theoretically, in more or less detail. It is now understood that λ^3-phosphorus ligands cannot only act as σ-donors and π-acceptors, but that their size and the nature of their substituents can also have an influence on their electronic properties as ligands. Probably the best method to determine the properties of phosphanes in great detail is the Quantitative Analysis of Ligand Effects (QALE) developed by Giering et al. Unfortunately, the method has not been employed to determine the ^{31}P-NMR chemical shift parameters. However, it stands to reason that the four QALE parameters exert a similar influence to the chemical shift in the NMR spectrum as they do to the νCO vibrations in the IR-spectrum.

It is interesting to note that good correlations between the phosphorus chemical shift and the P-M bond strength have only been found for phosphane-borane adducts where π-back bonding does not exist and steric factors are limited.

In IR-spectroscopy, the basic concept is that the νCO vibration is sensitive to the amount of electron density available on the metal, if a transition metal carbonyl complex like [M(CO)$_5$L] (M=Cr, Mo, W; L=phosphane, phosphite) is considered. We use [M(CO)$_5$L] rather than the more familiar Tolman system [Ni(CO)$_3$L] simply because ^{31}P-NMR data is more readily available for the former.

The position of the νCO bands in the IR-spectrum of a transition metal depends on the electronic situation at the metal. The CO ligand acts with its electron lone pair as a σ-donor toward the metal utilizing its HOMO, the antibonding 5 s (s_σ^*) orbital. The metal in turn can transfer electron density into the $2\pi(p_\pi^*)$ antibonding orbital (LUMO) of the CO ligand, weakening the CO bond and shifting the νCO band toward lower wave numbers. If a CO ligand is replaced by a tertiary phosphane, then the phosphane itself will act as a σ-donor with its electron lone pair towards the metal. This renders the metal more electron rich. This increased electron density on the metal can then be transferred toward the CO groups and the phosphane ligands. The extent of this back bonding is dependant on the π-acceptor strength of the various ligands. As only the phosphane is altered, and the other ligands (CO, Cp etc)

Box Story: (continued)

remain constant, the system can be used to measure the electronic properties of a series of phosphane ligands. Good π-acceptor properties are indicated by a shift of the vCO frequency toward higher wave numbers. However, as the shift of vCO frequency is only an indication for the net electron donating ability of the ligand, this is not sufficient proof for good π-acceptor strength.

In NMR terms, that means that the σ-donor ability of the phosphane transfers electron density from phosphorus to the metal. That results in a decrease of electron density on the phosphorus atom, and thus a downfield shift of the ^{31}P-NMR signal compared to the free ligand. The greater the σ-donor ability of the ligand, the larger the downfield shift. However, the π-acceptor strength operates in the opposite direction. Backbonding from the metal increases the electron density on phosphorus, and thus results in an upfield shift. The magnitude and sign of the net shift will be dependant on the phosphorus ligand, the metal, and the coligand.

Example 1 The pentacarbonyl group 6 complexes [M(CO)$_5$L] (M=Cr, Mo, W) contain a fairly early transition metal with limited backbonding ability, but five carbonyl ligands, very strong π-acceptors. We would expect that this metal centre readily accepts electron density from the phosphorus ligand, but backbonds instead to the carbonyl coligands. As a result, we expect a net downfield shift for poor to moderate π-acceptors like trialkyl and triaryl phosphanes that turns into a net upfield shift for good π-acceptor ligands like phosphites, and indeed that is what is observed (see Table 5.3).

Example 2 The complex *cis*-[PtCl$_2$L$_2$] contains a late transition metal that is more electron rich than the group 6 metals. Consequently, the PtCl$_2$-fragment appears to be a weaker σ-acceptor, but better π-donor than the W(CO)$_5$-fragment. We would therefore expect that the chemical shift values for *cis*-[PtCl$_2$L$_2$] are upfield from [W(CO)$_5$L], and that a net upfield shift is observed for weaker π-acceptor ligands in the platinum than in the tungsten series. Both assumptions are confirmed by experiment (see Table 5.3).

Table 5.3 Coordination chemical shifts for phosphanes and phosphites in group 6 and platinum complexes

Ligand	δ_p free ligand	[Cr(CO)$_5$L], $\Delta\delta$	[Mo(CO)$_5$L], $\Delta\delta$	[W(CO)$_5$L], $\Delta\delta$	*cis*-[PtL$_2$Cl$_2$], $\Delta\delta$
PMe$_3$	−62	+71	+45	+24	+37
PMe$_2$Ph	−48		+41		+32
PMePh$_2$	−28	+63	+43	+24	+27
PPh$_3$	−6	+61	+44	+27	+14
PBu^{i}Ph$_2$	+17	+55	+40	+24	
P(OPh)$_3$	+128		+17		−61
P(OMe)$_3$	+141	+39	+26	−4	−68
PCl$_3$	+218	−31	−66	−120	−102

Box Story: (continued)

Note: *In NMR, the influence of π-electrons on the position of the signal is significantly greater than that of σ-electrons. Therefore, the upfield shift due to π-backbonding can overcompensate the downfield shift due to the σ-donicity of the ligand, even though the actual bonding contribution is in reverse order.*

We need to remember that σ-donor and π-acceptor properties of the phosphorus ligands are not the only factors that contribute to the value of the vCO vibration. In fact, there are four QALE parameters:

$$\text{property} = a\,\chi_\text{d} + b\,\Theta + c\,E_\text{ar} + d\,\pi_\text{p} + e$$

with χ_d : σ-donor capacity; Θ : Tolman's cone angle; E_ar : aryl effect parameter; π_p : π-acidity, and they stand for the σ-donor and π-acceptor strength, the steric threshold value, and the contribution from the substituents on phosphorus. It is safe to assume that not only the first two, but also the last two have an influence on the phosphorus chemical shift. This is most obvious for the steric threshold value, since steric constraints will inevitably lead to changes in the bond angles around phosphorus, and the phosphorus bond angles are a known contributory factor in changes of the phosphorus resonance.

The steric contribution is usually expressed in terms of the Tolman cone angle. The Tolman cone angle is defined as the angle the substituents on phosphorus form with the metal atom in [Ni(CO)$_3$L] using the fragment MPR$_3$ and a standard M-P bond length of 228 pm. Where the substituents on phosphorus are not identical, an average value is used. Experimental data on the series *cis*-[M(CO)$_4$(PPh$_2$CH$_2$PR$_2$)] (M = Cr, Mo, W; R = Me, Pr$^\text{i}$, Ph) shows only a small effect that is not consistent going down the group (see Table 5.4).

The question we have to ask ourselves while looking at the data compiled in Table 5.4 is this: Is this system a valid model for the monitoring of the cone angle influence, and, in particular, did we select the right substituents? The answer is clearly no. By changing the substituent R from alkyl to aryl and back to alkyl, we in fact change the ligand class of the phosphane and thus introduce electronic influences that obscure the steric influence that we want to measure.

In fact, in Sect. 7.1 we will encounter a set of equations with which we can calculate the chemical shift of the coordinated phosphorus ligand

Table 5.4 Dependance of the coordination chemical shift on the cone angle Θ

R	$\Theta[°]$	$\Delta\delta$ [ppm] Cr	$\Delta\delta$ [ppm] Mo	$\Delta\delta$ [ppm] W
Me$_2$	101	+57.5	+27.3	−1.4
Me, Ph	110	+52.9	+26.1	−0.5
Ph$_2$	119	+49.0	+23.6	0.0
Pr$^\text{i}$, Ph	124	+48.3	+25.9	+2.0
Pr$^\text{i}_2$	129	+48.1	+28.0	+5.4

Box Story: (continued)

from the chemical shift of the free ligand. Since those equations are based on electronic factors, they can only be valid if there is no steric influence. The QALE method tells us to expect a steric influence only for very bulky ligands for which a steric threshold value is reached. If this is true, then significant deviations from the calculated values should be observed for small ligands and for bulky ligands, but with opposite signs.

5.2 Phosphorus Carbon Multiple Bonds

Turning our attention to molecules with carbon phosphorus multiple bonds, we acknowledge the existence of π-bonding between carbon and phosphorus in these molecules. Since carbon is more electronegative than phosphorus, we would suspect that a carbon phosphorus multiple bond would result in a downfield shift of the phosphorus resonance. Indeed, the ^{31}P-NMR spectra of simple phosphaalkenes usually show a resonance of $\delta_p = 200$–300 ppm.

Going from a phosphaalkene to a phosphaalkyne, we increase the π-contribution in the carbon phosphorus multiple bond, and would therefore expect a further downfield shift of the phosphorus resonance. However, a glance at the situation in carbon carbon multiple bond systems; in particular, alkenes and alkynes, tells us that ^{13}C-NMR spectra of these molecules show the carbon resonance of alkynes upfield from that of alkenes. This is usually explained by anisotropic effects associated with the linear rod-shaped structure of alkynes versus the bend structure of alkenes. As the geometries of phosphaalkenes and phosphaalkynes are analogous to alkenes and alkynes, respectively, we can assume that the explanation given for the appearance of the carbon resonance in alkynes upfield from that for alkenes in ^{13}C-NMR spectra is also applicable for the respective unsaturated phosphorus compounds.

The resonance of the phosphorus signal in the ^{31}P-NMR spectrum can be shifted downfield or upfield by changing the substituent on the carbon atom of the carbon phosphorus triple bond. Table 5.5 shows the influence of the *para*-substituent on the aromatic ring of a series of Ar—C≡P compounds. As the substituent operates

Table 5.5 ^{31}P chemical shifts of selected phosphaalkynes

R	δ_p [ppm]	$^1J_{PC}$ [Hz]
H	37.2	54.0
OMe	32.4	52.4
NMe$_2$	32.1	49.8
Bu^t	34.4	53.2

Fig. 5.10 ^{31}P chemical shifts for a series of phosphaalkynes

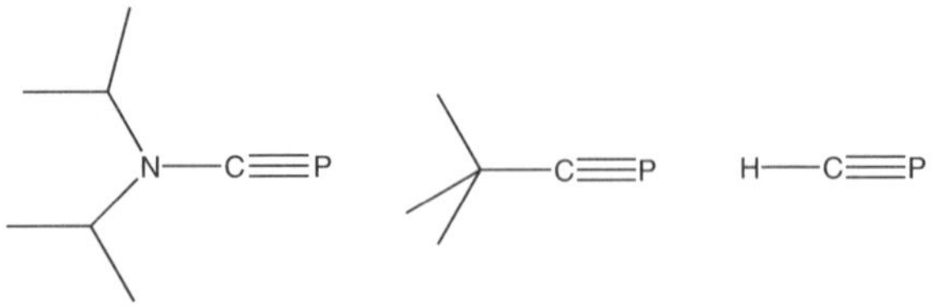

δ_P [ppm]	−99.6	−69.2	−32
$^1J_{PC}$ [Hz]	14.7	38.5	56.0

on the aromatic ring and through the aromatic ring on the phosphorus atom, we would expect an increasing upfield shift for the phosphorus resonance in the order H<alkyl<alkoxy<amino. The experimentally observed changes are small, but follow exactly the predicted sequence.

Note: *The $^1J_{PC}$ coupling constants decrease in line with increasing upfield shift of the phosphorus resonance, indicating decreasing s-character of the phosphorus lone pair.*

If we take away the aryl ring acting as a mediator between the substituent and the carbon atom of the carbon phosphorus triple bond, we would still expect the same shift of the phosphorus resonance qualitatively, but of course to a larger degree quantitatively. Remember, the substituent still acts on the carbon atom of the triple bond, and not the phosphorus atom. This is illustrated in Fig. 5.10, where we see a significant upfield shift of the phosphorus resonance in the order H<alkyl<amino. The individual shift differences are approximately $\Delta\delta=-33$ ppm, and thus one order of magnitude larger than for the para-substituted arylated phosphaalkynes in Table 5.5.

Note: *The decrease in the $^1J_{PC}$ coupling constants has likewise increased in magnitude.*

In Table 5.6 , we see this substituent effect apparently in reverse order. However, the aryl ring is now bonded to the phosphorus atom of a phosphaalkene. This has the effect that an increase of electron density on this aryl substituent of phosphorus will ultimately increase the electron density on phosphorus through the σ-bond in the accustomed order H<alkyl<alkoxy<amino. The newly electron enriched

Table 5.6 ^{31}P chemical shifts for a series of phosphaalkenes

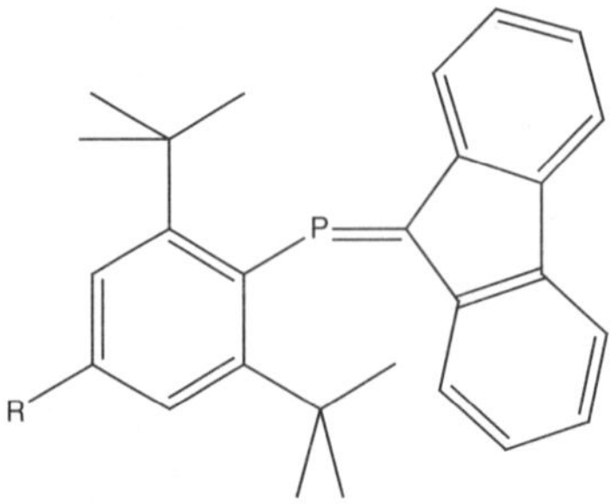

	R	δ_P [ppm]	R	δ_P [ppm]
	H	254.8	NMe$_3$/I	238.3
	Bu^t	256.5	NMe$_3$/OTf	238.8
	OMe	258.0		
	NMe$_2$	264.8		

phosphorus atom passes part of this electron density increase on to the other carbon substituent using the π-bond. This cannot be done directly, of course, since the two π-systems are orthogonal to each other. Instead, the π-donor destabilises the phosphorus lone pair with a subsequent decrease in the $\Delta E\, n_p \rightarrow \pi^*_{P=C}$ contribution, resulting in deshielding.

Note: *Protonation of the amino group results in an upfield shift as the electron density on phosphorus is decreased through the σ-bond, resulting in a shift of π-density toward phosphorus, and thus a net upfield shift of the phosphorus resonance.*

The phosphorus chemical shift of Mes-P=CPh$_2$ is $\delta_p = 233.1$ ppm upfield from the resonance observed for molecules shown in Table 5.6, where the extended delocalized π-electron system of the fluorenyl moiety can distribute partial charges significantly better than the isolated phenyl rings in Mes-P=CPh$_2$.

In Table 5.7, we again see an example for the dependance of the phosphorus resonance on the electronic changes within the substituents on phosphorus. As the electron density on the aromatic substituents R is increased (from methyl to methoxy), the two phosphorus resonances are shifted upfield. However, if R=Bu^t, then both phosphorus resonances are observed upfield, although in the simple phosphanes PBu$^t_3$ and P(p-tol)$_3$, the chemical shift for PBu$^t_3$ is observed downfield from P(p-tol)$_3$.

We note that the P=C—P unit is a 1,3-diphosphavinyl system giving the phosphorus lone pair the opportunity to be in conjugation with the phosphaalkene double bond. For effective conjugation to occur, the substituents on the tertiary phosphorus atom have to be coplanar with the phosphaalkene unit. However, hyperconjugation operates when the dihedral angle is different from 90° and is angle dependent. We observe that the effect is both steric and electronic in nature, and operates on both phosphous atoms in parallel.

Table 5.7 Dependance of the phosphorus chemical shift on the substituents of the phosphino group in a series of phosphino phosphaalkenes

R	δ_p [ppm]	δ_p [ppm]	$^2J_{PP}$ [Hz]
Bu^t	277.0 d	– 5.4 d	214
p-MeC$_6$H$_4$	289.0 d	8.5 d	243
p-MeOC$_6$H$_4$	284.8 d	7.8 d	214

Table 5.8 Oxidation of the phosphino phosphorus atom in a series of phosphino phosphaalkenes

R	δ_p [ppm]	δ_p [ppm]	$^2J_{PP}$ [Hz]
Ph	320.9 d	33.0 d	115
Bu^n	309.8 d	47.0 d	92
p-MeC$_6$H$_4$	318.2 d	33.2 d	116
p-MeOC$_6$H$_4$	316.8 d	32.6 d	116

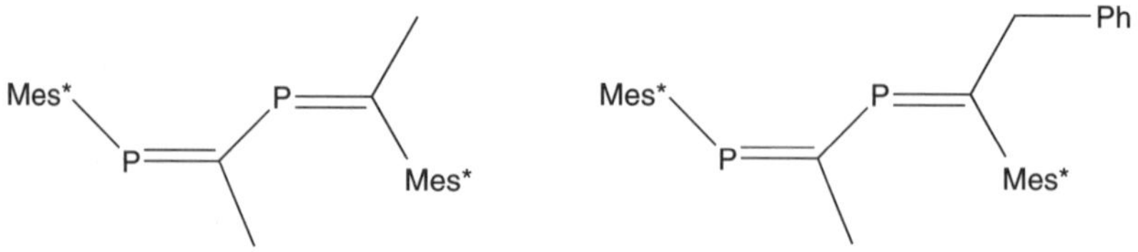

δ_P: 318.3 ppm dqq   237.9 ppm dq   326.0 ppm dqt   233.7 ppm dt

$^2J_{PP}$:   29 Hz   29 Hz   26 Hz   26 Hz

$^3J_{PH}$:   27 Hz   24 Hz   28 Hz   19 Hz

$^5J_{PH}$:   5 Hz   7 Hz

Fig. 5.11 The ^{31}P chemical shifts for selected 1,3-diphosphabutadienes

If the tertiary phosphane is oxidized to the phosphanoxide, we observe down-field shifts for both phosphorus resonances. That is expected, as phosphanoxides resonate at $\delta_p = 10$–70 ppm, and the diphosphavinyl system is broken up. There is a diminished influence of the *p*-substituent on the aryl group R.

In Fig. 5.11, we see a 1,3-diphosphabutadiene system and need to familiarise ourselves with the coupling pattern. The conjugated π-electron system enables the outer phosphorus atom to couple both to the central methyl group (allyl position, $^3J_{PH}$ coupling) and the outer alkyl group ($^5J_{PH}$ coupling), whereas the inner phosphorus atom can only couple to the outer alkyl group (allyl position, $^3J_{PH}$ coupling).

When the outer methyl group is substituted by the more electron withdrawing benzyl group, the outer carbon atom of the diphosphabutadiene backbone loses electron density through the σ-bond, and thus draws more electron density from the inner phosphorus atom that ultimately regains its loss from the outer phosphorus atom. This explains the downfield shift of the resonance for the outer phosphorus atom. For the inner phosphorus atom, the gain occurs via the π-electron network, whereas the loss to the outer carbon atom seems to be predominantly σ in nature, leading to an overall upfield shift.

The unsaturated phosphorus compounds depicted in Fig. 5.12 show a quinoidal structure, and resonate in the normal range for a P=C double bond. The thienyl derivative has an electron-rich heterocycle (thiophene) as parent compound. The

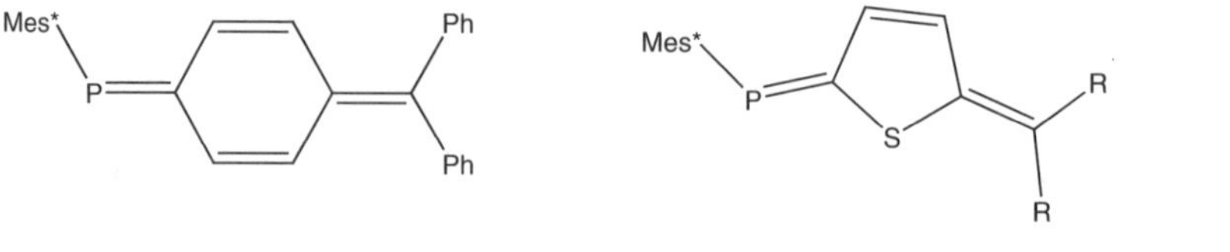

R: Ph   *p*-MeC$_6$H$_4$ *p*-MeOC$_6$H$_4$

δ_P: 244.4 ppm   201.4 ppm   198.9 ppm   196.3 ppm

Fig. 5.12 The influence of a quinoidal system on the phosphorus resonance of selected phosphaalkenes

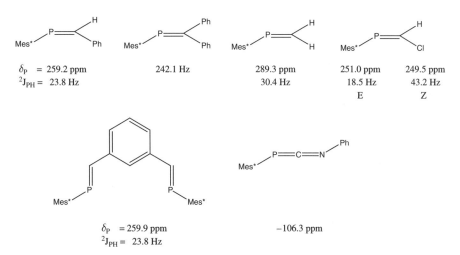

δ_P = 259.2 ppm   242.1 Hz   289.3 ppm   251.0 ppm   249.5 ppm
$^2J_{PH}$ = 23.8 Hz   30.4 Hz   18.5 Hz   43.2 Hz
                                                               E   Z

δ_P = 259.9 ppm   −106.3 ppm
$^2J_{PH}$ = 23.8 Hz

Fig. 5.13 Dependance of the ^{31}P chemical shift on the phosphorus substituents in a series of selected phosphaalkenes

added electron density emanating from the sulfur lone pair moves the phosphorus chemical shift to a higher field compared to the six-membered ring system in the phospha-benzoquinone. The influence of the heteroatom in the five- membered heterocycles, such as thiophene, pyrrole, or furane, on the phosphorus chemical shift is frequently seen in ^{31}P-NMR spectroscopy.

In Fig. 5.13, we see the influence substituents on the carbon atom of the carbon phosphorus double bond exert on the phosphorus resonance. The least electro-negative substituent, hydrogen, effects the greatest downfield shift, $\delta_p = 289.3$ ppm. Replacing the hydrogen atom with a phenyl group or a chlorine atom shifts the phosphorus resonance upfield by $\Delta\delta = -29.9$ and -39.8 ppm, respectively. This behavior must be related to the ability of the π-electron system of the phenyl group

δ_P: 313.8 ppm   321.5 ppm
      291.3 ppm
      264.7 ppm   292.9 ppm
                                                           263.9 ppm

Fig. 5.14 Phosphorus resonances of selected phosphafulvene and phosphole compounds

or the chlorine lone pairs to interact with the π-electrons of the P=C double bond, as the σ-effects are dominated by electronegativity considerations.

The introduction of a second phenyl group increases the upfield shift of the phosphorus resonance to $\delta_p = 242.1$ ppm, corroborating the findings. If the phenyl substituent serves two phosphaalkene moieties, the phosphorus chemical shift is at the same position as in Mes*-P=CHPh, $\delta_p = 259.9$ ppm.

The heteroallene system in Mes*-P=C=N-Ph gives rise to an upfield shifted phosphorus signal at $\delta_p = -106.3$ ppm, similar to that in Pr^i_2N—C≡P, $\delta_p = -99.6$ ppm, and doubtless for the same reasons.

The phosphorus resonances in 1,3,6-triphosphafulvene compare well with those in 2,4,6-tri-*tert*-butyl-1,3,5-triphosphabenzene. In both compounds, the phosphorus replaces an isoelectronic CR-unit. The phosphorus resonance in 2,4,6-tri-*tert*-butyl-1,3,5-triphosphabenzene at $\delta_p = 232.6$ ppm is some $\Delta\delta = -30$ to -60 ppm upfield from that in 1,3,6-triphosphafulvene. The corresponding Dewar benzene analogue depicted in Fig. 5.15 is no longer planar, and features two distinctly different phosphorus atoms, a trivalent atom with a chemical shift of $\delta_p = 93.6$ ppm, and a divalent atom (with an isolated P=C double bond) at $\delta_p = 336.8$ ppm, similar to the resonance of the exocyclic phosphorus atom in 1,3,6-triphosphafulvene.

In the Dewar-type isomer of 2,4,6-tri-*tert*-butyl-1,3,5-triphosphabenzene, the P=C double bonds are isolated, whereas in 1,3,6-triphosphafulvene, they are conjugated, and in 2,4,6-tri-*tert*-butyl-1,3,5-triphosphabenzene, they are embedded in a heteroaromatic ring system. The trend is obvious, as the degree of conjugation increases the phosphorus chemical shift moves upfield. Comparison of the phosphorus resonances of the exocyclic phosphorus atom in 1,3,6-triphosphafulvene with those in phosphafulvene and the Dewar-type 2,4,6-tri-*tert*-butyl-1,3,5-triphosphabenzene shows that the diphosphacyclopentadiene moiety has only a small effect on the position of the signal for the exocyclic phosphorus atom in 1,3,6-triphosphafulvene.

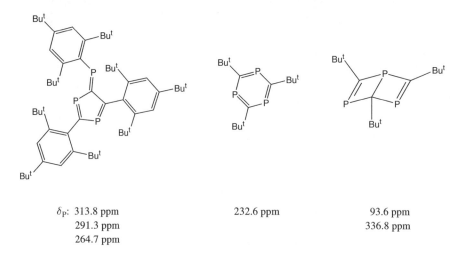

δ_P: 313.8 ppm   232.6 ppm   93.6 ppm
    291.3 ppm   336.8 ppm
    264.7 ppm

Fig. 5.15 Comparison between the phosphafulvene and phosphabenzene systems

Lithiation with subsequent alkylation results in alkylation of the exocyclic phosphorus atom, and formation of an aromatic system in the central heterocycle analogous to a cyclopentadienide. Spectroscopically, the effect is a severing of the exocyclic phosphorus atom from the conjugated ring system, accompanied by a significant upfield shift of the phosphorus resonance by over $\Delta\delta = -300$ ppm, dependent on the alkyl substituent.

The effect on the endocyclic phosphorus atoms is less pronounced, as they stay within the conjugated double bond system, but since a negative charge and thus more electron density is introduced, the phosphorus resonance for these two phosphorus atoms is shifted upfield as well, by $\Delta\delta = -48$ and $\Delta\delta = -68$ ppm, respectively, independent of the alkyl group.

Subsequent alkylation of one of the endocyclic phosphorus atoms results in loss of the aromatic ring system and formation of a cyclic phosphadiene. The implications to the phosphorus resonances are a significant upfield shift for the alkylated endocyclic phosphorus atom as it is released from the π-system, and a downfield shift of $\Delta\delta = 50$ ppm of the resonance for the remaining endocyclic double bonded phosphorus due to the removal of negative charge from the conjugated π-electron system. The resonance for the exocyclic phosphorus atom is almost unchanged.

If the 1,3,6-triphosphafulvene is converted into an ylide by oxidation of the exocyclic phosphorus atom, a small downfield shift of all three phosphorus resonances is observed. Although it is a formal oxidation, it is carried out as a protonation of the anionic compound depicted in Fig. 5.16. We can look at it as a protonation of the lone pair on the exocyclic phosphorus atom and subsequent formation of an intramolecular donor bond from the endocyclic carbon atom to the exocyclic phosphorus atom. The 1,3,6-triphosphafulvene system is reformed, but now with two additional substituents on the exocyclic phosphorus atom.

The resonance for the exocyclic phosphorus atom is found at $\delta_p = -23.2$ ppm for R=Me in the region customary for ylides, and is shifted to lower field, $\delta_p = -4.5$ ppm, for R=Bu^n as expected. For R=F, the resonance shifts even further downfield, $\delta_p = 45.2$ ppm, in line with the greater electronegativity of fluorine compared to carbon.

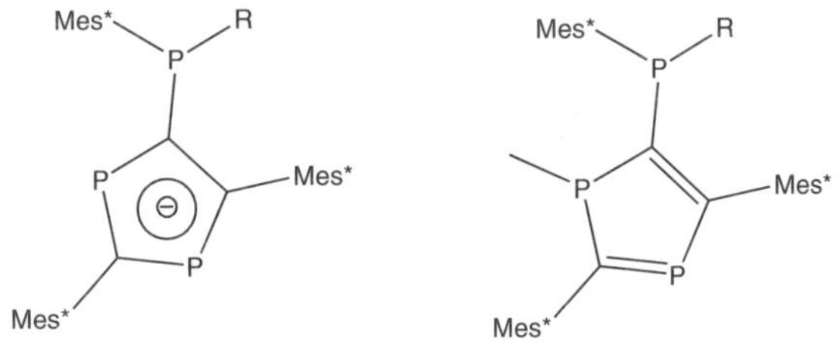

	R :	Me	Bu^n	Me	Bu^n
Fig. 5.16 Lithiated and alkylated 1,3,6- triphosphafulvene systems	δ_p:	223 ppm	223 ppm	272 ppm	265 ppm
		216 ppm	216 ppm	66 ppm	67 ppm
		−31 ppm	−2 ppm	−21 ppm	−8 ppm

Fig. 5.17 The phosphorus resonances in phosphafulvene-based phosphorus ylides

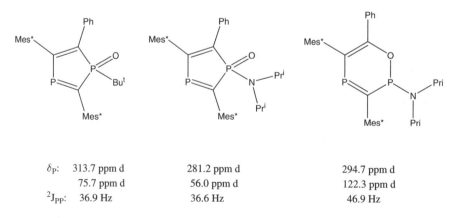

R :	Me		Bu^n	
δ_P:	233.7 ppm d	240.0 ppm ddm	249.4 ppm dpt	
	229.3 ppm ddd	235.6 ppm d	247.0 ppm ddd	
	−23.2 ppm ddq	−4.5 ppm ddm	45.2 ppm dddd	

The effect on the endocyclic phosphorus atoms is only small, but downfield ($\Delta\delta = 10.7$ and $\Delta\delta = 13.3$ ppm for R=Me, and $\Delta\delta = 17.0$ and $\Delta\delta = 19.6$ ppm for R=Bu^n, respectively) and larger for R=F ($\Delta\delta = 26.4$ and 31.0 ppm, respectively). Comparison to the endocyclic phosphadiene system in Fig. 5.16 shows a significant upfield chemical shift of $\Delta\delta = -38.3$ and -42.7 ppm for R=Me, and $\Delta\delta = -25.0$ and -29.4 ppm for R=Bu^n, respectively. No direct comparison for R=F is possible. The endocyclic phosphorus atoms are also shifted upfield from the values for the 1,3,6-triphosphafulvene parent compound.

Oxidation of the exocyclic phosphorus atom of the 1,3,6-triphosphafulvene system disturbs the conjugation and shifts all three phosphorus resonances upfield.

Phosphanoxides usually resonate downfield from ylides, and it is therefore not surprising that the phosphorus resonances for the phosphanoxides in Fig. 5.18 are observed at $\delta_P = 75.7$ and 56.0 ppm, respectively. The influence of the amino group is seen in an upfield shift of $\Delta\delta = -19.7$ ppm. The λ^3-phosphorus atom resonates significantly downfield at $\delta_P = 313.7$ ppm, and thus in the region of the exocyclic

δ_P:	313.7 ppm d	281.2 ppm d	294.7 ppm d
	75.7 ppm d	56.0 ppm d	122.3 ppm d
$^2J_{PP}$:	36.9 Hz	36.6 Hz	46.9 Hz

Fig. 5.18 ^{31}P chemical shifts for P-oxidized phosphafulvenes

phosphorus atom of the 1,3,6-triphosphafulvene. The amino group moves that resonance upfield by $\Delta\delta = -32.5$ ppm.

In the six-membered heterocycle, the λ^3-phosphorus atom resonates at
$\delta_p = 122.3$ ppm, and thus considerably downfield from the λ^5-phosphorus atoms of
the other two compounds. Here, the phosphorus atom is in allyl position toward the
phosphabutadiene system and thus participates with its lone pair in the conjugated
π-electron system, resulting in the observed downfield shift.

An interesting class of compounds is the motif displayed in Fig. 5.19, where two
phosphorus atoms sit in the 1 and 3 positions of a cyclobutane system, with the two carbon atoms each formally possessing an unpaired electron. Nonetheless, the molecules
are diamagnetic and show phosphorus resonances that are characteristic of phosphanes
rather than phosphinidenes.

Theoretical calculations show that the diradical form is usually more stable than
the equally possible λ^5-phosphane species seen at the far right of Fig. 5.19. Structures with a carbon carbon single bond forming a bicyclic compound and aromatic
6 π-electron system involving both phosphorus lone pairs are predicted to be even
less stable.

Spectroscopically, the difference between the structure with λ^3- and λ^5-phosphorus
centres can clearly be seen from the ^{31}P-NMR chemical shift values that are in the
range between $\delta_p = -20$ and 80 ppm for the λ^3-phosphane, and about $\delta_p = 122$ ppm
for the single λ^5-phosphane centre depicted in Fig. 5.19. The difference in the two
phosphorus resonances becomes even more apparent when the substituents on
phosphorus are taken into consideration. As the second phosphorus atom is oxi-
dised to the phosphanoxid, the phosphorus resonance is shifted downfield from
$\delta_p = -18.4$ ppm to $\delta_p = 121.8$ ppm, a difference of $\Delta\delta = 140.2$ ppm in accordance
with the formation of localized π-bonds on the phosphorus atom bearing the amino
substituent. Since this amino substituent is likely to act as a π-donor toward the
phosphorus atom, the considerable upfield shift of the phosphorus resonance of
$\Delta\delta = -128$ ppm relative to the normal value for a P=C double bond of $\delta_p = 250$ ppm
seems credible.

Note: *The two isomers in Fig. 5.19 can easily be distinguished by their ^{31}P-NMR
chemical shift values.*

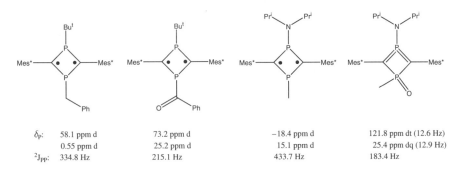

δ_P:	58.1 ppm d	73.2 ppm d	−18.4 ppm d	121.8 ppm dt (12.6 Hz)
	0.55 ppm d	25.2 ppm d	15.1 ppm d	25.4 ppm dq (12.9 Hz)
$^2J_{PP}$:	334.8 Hz	215.1 Hz	433.7 Hz	183.4 Hz

Fig. 5.19 A series of 1,3-diphospha-cyclobutadiene-based diradicals

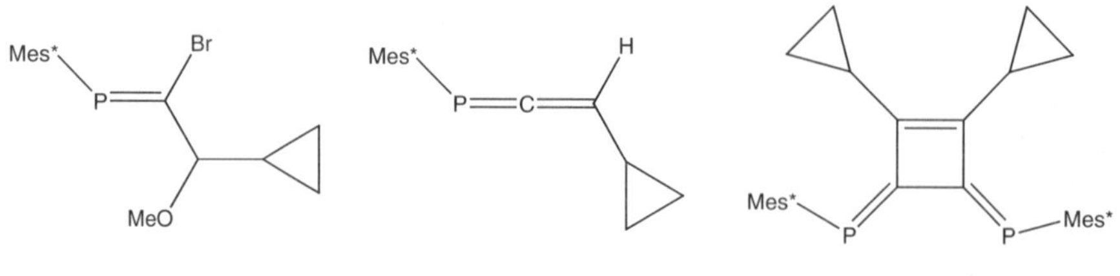

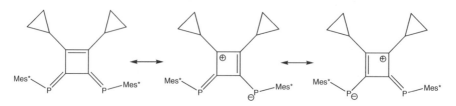

δ_P: 259.6 ppm   70.6 ppm   147.0 ppm

Fig. 5.20 Comparison between the ³¹P chemical shifts of isolated and conjugated P═C bonds

In Fig. 5.20,we see a typical representative of a phosphaalkene with a ³¹P-NMR chemical shift value of $\delta_p = 259.6$ ppm, and a phosphaallene with an equally typical upfield resonance of $\delta_p = 70.6$ ppm. The relative chemical shift values of these two compound classes comprising P═C double bonds mirror closely the ¹³C-NMR chemical shift values for olefins and acetylenes, with the signals for the acetylenes being observed upfield from the olefine resonances.

The ³¹P-NMR chemical shift for the 3,4-diphosphinidene-cyclobutene depicted on the right in Fig. 5.20 is observed at $\delta = 147$ ppm shifted significantly upfield from the normal value for P═C double bonds. The presence of two electron-rich cyclopropyl substituents on the cyclobutene ring cannot explain this phenomenon. The cyclopropyl substituents are, however, capable to effectively stabilise a resonance structure featuring a carbenium cation and a phosphanyl anion. Such a resonance structure is depicted in Fig. 5.21, and is possible for either phosphorus atom. In the charge separated resonance structures featuring a carbenium cation and a phosphanyl anion, the phosphorus atom acquires a formal negative charge while no longer engaged in a π-interaction with the ring carbon atom. Both of these properties translate into the significant upfield shift actually observed for this compound class.

Corroborating evidence comes from crystal structures reported for 3,4-diphosphinidene-cyclobutene, in which the C(cyclopropyl)-C(cyclopropyl) bonds are longer, and the C(cyclopropyl)-C(P) bonds are shorter than those expected for C═C double and C—C single bonds, respectively.

It is interesting to note that the benzodiphospholide shown in Fig. 5.22 with a phosphorus resonance at $\delta_p = 153.4$ ppm is almost identical to the ³¹P-NMR chemical

Fig. 5.21 Resonance structures in 3,4-diphosphinidene-cyclobutene

Fig. 5.22 Comparison of the structures of 3,4-diphosphinidene-cyclobutene with benzannelated phospholes

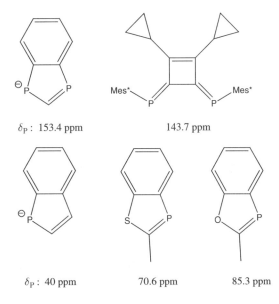

δ_P : 153.4 ppm   143.7 ppm

δ_P : 40 ppm   70.6 ppm   85.3 ppm

shift value of $\delta_p = 147$ ppm observed for the 3,4-diphosphinidene-cyclobutene, since the only thing in common between the two structures is a delocalized π-electron system of different extent.

Note: *It is evident from the very similar phosphorus resonances of the two compounds that the embeddedness into a heteroaromatic ring system as opposed to a conjugated "acyclic" system can still result in a similar phosphorus resonance.*

Note: *The three compounds in the lower half of Fig. 5.22 show the phosphorus resonance in phosphaindolyl at $\delta_p = 40$ ppm, $\Delta\delta = -107$ ppm upfield to 3,4-diphosphinidene-cyclobutene, highlighting the absence of a P═C double bond. Substitution of the CH- unit for an isoelectronic sulfur or oxygen atom results in the expected downfield shift of the phosphorus resonance of $\Delta\delta = 30.6$ ppm and 45.3 ppm, respectively. These downfield shifts are indicative of the increase in electronegativity in the order $C < S < O$.*

Note: *P^-, S and O all can engage in π-donor interactions diminishing the C═C and C═P bond order. This explains the significant upfield shift of the latter two phosphorus resonances and downfield shift in the former.*

Benzazaphospholes show a similar irregularity in their ^{31}P-NMR chemical shift values. Their phosphorus resonances depend on the position of the heteroatoms within the aromatic rings, as well as the substituents. In Table 5.9, a series of benzazaphospholes is shown where the phosphorus atom is in 3-position to nitrogen, and thus in conjugation with the nitrogen lone pair. The effect is a pronounced upfield shift of the phosphorus resonance to a value of $\delta_p = 63.5–80.5$ ppm. The influence of the substituents R^1, R^2 and R^3 is in line with the mesomeric and inductive effects usually observed with these substituents on aromatic or heteroaromatic ring systems.

Table 5.9 Influence of substitution in benzazaphospholes

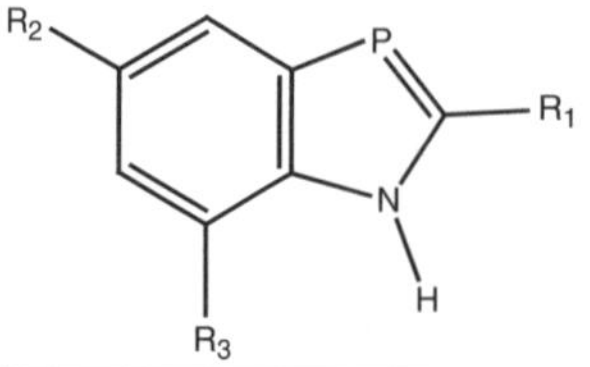

	R$_1$	R$_2$	R$_3$	δ_P [ppm]	$^1J_{PC}$ [Hz]
	Me	H	H	69.8	
	Me	Me	H	71.8	51.5
	Bu^t	Me	H	63.5	57.5
	Ph	Me	H	74.4	51.3
	Me	F	F	80.5	54.5

Metallation of the benzazaphosphole results in a very moderate upfield shift of $\Delta\delta < -10$ ppm. This can easily be explained by looking at the electronic structures of the metallated and non-metallated benzazaphospholes seen in Fig. 5.23. Only one of the resonance structures for the metallated benzazaphosphole is shown, but it is obvious that the negative charge is likely to reside on nitrogen, with at least equal probability making the situation for the phosphorus atom not so much dissimilar from that in the non-metallated benzazaphosphole.

Not surprisingly, when the neutral benzazaphosphole is coordinated to a metal fragment like [M(CO)$_5$] (M = Cr, Mo, W), the upfield shift for the lithiated complex, relative to the protonated one, is markedly larger than in the free ligand.

When the nitrogen atom lone pair of a benzazaphosphole is prevented from being in conjugation with the phosphorus atom, either by its relative position in the ring system or influences from the substituents, the phosphorus resonance is significantly downfield and near the usual value for a P=C double bond. Examples are shown in Fig. 5.24.

The influence of an additional lone pair in a heteroallene structure can be seen in Fig. 5.25. A typical P=C=C system is shown at the bottom with a chemical shift of $\delta_P = 70.6$ ppm. The P=C=N system at the top is considerably upfield shifted at $\delta_P = -106.3$ ppm. The reason is the additional lone pair on nitrogen that can form a π-bonding interaction with the neighbouring carbon atom, resulting in the removal of the P=C double bond and a negative charge on phosphorus. The difference between a P—C single and a P=C double bond in ^{31}P-NMR terms is usually $\Delta\delta = 200$–300 ppm, indicating that the resonance structure on the right is indeed relevant for phosphorus NMR purposes.

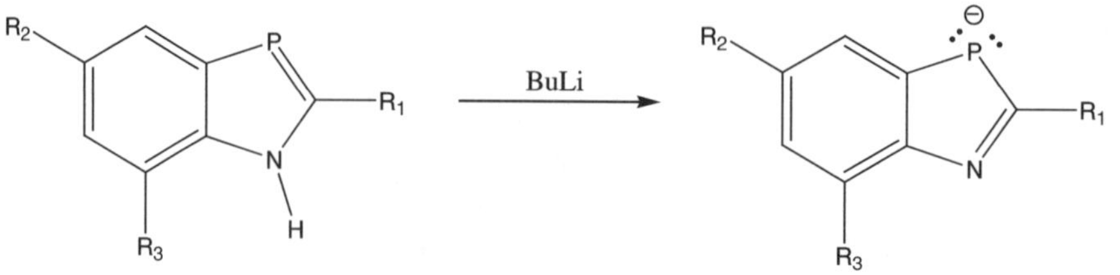

Fig. 5.23 Influence of lithiation in the phosphorus chemical shifts of benzazaphospholes

Fig. 5.24 Influence of the
nitrogen atom position on
the ^{31}P chemical shift in
benzazaphospholes

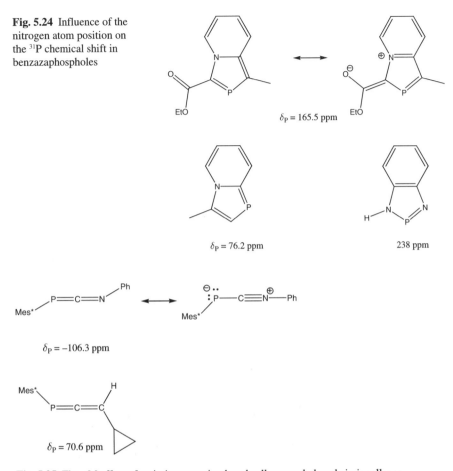

δ_P = 165.5 ppm

δ_P = 76.2 ppm

238 ppm

δ_P = −106.3 ppm

δ_P = 70.6 ppm

Fig. 5.25 The +M effect of an iminogroup in phosphaallenes and phosphaiminoallenes

5.3 Phosphorus Phosphorus Bonds

In the previous two subchapters, we have seen the chemical shift values for P—C single and P═C double bonds and the factors that influence each. With this knowledge, we should be able to deduce the relative chemical shift ranges for P—P single and P═P double bonds, assuming that similar factors apply as for the PC compounds.

We remember that one of the most important factors contributing to the position of the phosphorus resonance of a phosphane is the electronegativity of its substituents. We found that carbon, being more electronegative than hydrogen, causes a downfield shift as one goes from monophosphane PH$_3$ over primary and secondary to tertiary phosphanes. Since phosphorus is less electronegative than carbon – in fact about as electronegative as hydrogen – we would expect an upfield shift upon substitution of an alkyl or aryl group with a phosphino group, if the chemical shift were dependant on electronegativity alone. Seemingly, the other contributing factors are dominant here.

Similarly, we would expect the diphosphenes RP=PR to be slightly upfield from the phosphaalkenes. We will see that this assumption is not valid. In fact, diphosphenes resonate considerably downfield from their corresponding phosphaalkenes.

A very interesting class of phosphorus compounds are the phospholes and phospholides, of which we have already encountered some representatives in Sect. 5.2. There are some important members of this family with P—P bonds, and we will study these compounds in more detail in this chapter.

5.3.1 Diphosphanes

Diphosphanes consist of two PR_2 units fused together by a P-P single bond. They can be regarded as a phosphane where a carbon substituent is exchanged for a second phosphino group. They can be symmetrically or unsymmetrically substituted, and they are frequently chiral, if they carry two different groups on either or both phosphorus atoms (see Fig. 5.26). In the latter case, there are two chiral centres allowing for the existence of diastereomers.

From the relative electronegativities, we assume that a diphosphane resonates upfield from the corresponding phosphane. A selection of diphosphanes is depicted in Fig. 5.27, together with the chemical shift values. It should be mentioned that we have a choice as to which is the corresponding phosphane. With other words, do we substitute a methyl, tert-butyl, or phenyl group? A significant decision, if we look at the phosphorus resonances of an eligible tertiary phosphane series, e.g. PPh_2Me ($\delta_p = -28$ ppm), PPh_3 ($\delta_p = -6$ ppm), PPh_2Bu^t ($\delta_p = 16$ ppm), a difference of $\Delta\delta = 44$ ppm going from methyl to tert-butyl.

From the first three entries in Fig. 5.27, we see that there is no simple quantitative relationship between the $\Delta\delta$ value and the substituents on the PR_2 group. The three diphosphanes: P_2Me_4, P_2Ph_4 and Me_2PPPh_2 have phosphorus resonances of $\delta_p = -58.5$ ppm, $\delta_p = -14.7$ ppm, $\delta_p = -64.2$ ppm (PMe_2), and $\delta_p = -11.5$ ppm (PPh_2), respectively. Thus, P_2Me_4 resonates downfield from PMe_3, and P_2Ph_4 upfield from PPh_3. The really revealing compound, however, is Me_2PPPh_2, where the PMe_2 group shifts the phosphorus resonance of the PPh_2 group downfield compared to PPh_2 in P_2Ph_4 ($\Delta\delta = 3.2$ ppm), and the chemical shift of the PMe_2 group is moved upfield by PPh_2 compared to P_2Me_4 ($\Delta\delta = -5.7$ ppm). We suspect that the two phosphino

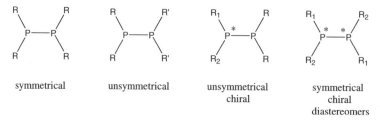

| symmetrical | unsymmetrical | unsymmetrical chiral | symmetrical chiral diastereomers |

Fig. 5.26 Possible substitution patterns on diphosphanes

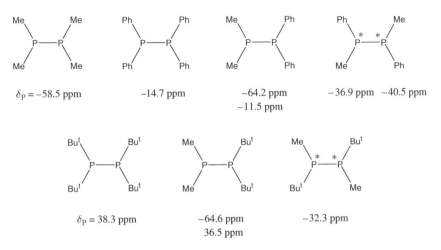

$\delta_P = -58.5$ ppm -14.7 ppm -64.2 ppm -36.9 ppm -40.5 ppm
-11.5 ppm

$\delta_P = 38.3$ ppm -64.6 ppm -32.3 ppm
36.5 ppm

Fig. 5.27 ^{31}P chemical shift values for some selected diphosphanes

groups mutually influence each other, possibly through the two phosphorus atom lone pairs.

However, we can deduce from Fig. 5.27 that the phosphino groups of a diphosphane usually resonate slightly upfield from their "parent phosphanes." This is particularly pronounced in the case of *tert*-butyl groups. Whereas PBu$^t_3$ has a chemical shift value of $\delta_P = 63$ ppm, the corresponding diphosphane P$_2$Bu$^t_4$ resonates at $\delta_P = 38.3$ ppm, and thus $\Delta\delta = -24.7$ ppm upfield.

Figure 5.28 shows the influence of a hydrogen substituent and that of the trifluoromethyl group on the chemical shift values of the diphosphanes. Whereas HCF$_3$PPHCF$_3$ has two chiral phosphorus centers resulting in two diastereomers with phosphorus chemical shifts of $\delta_P = -90.3$ ppm and $\delta_P = -92.0$ ppm, respectively, the unsymmetrical diphosphane Me$_2$PP(CF$_3$)$_2$ has phosphorus chemical shifts of $\delta_P = 10.8$ ppm for the P(CF$_3$)$_2$ group and $\delta_P = -57.5$ ppm for the PMe$_2$ group.

Primary phosphanes usually resonate upfield from HCF$_3$PPHCF$_3$, whereas secondary phosphanes are seen downfield. The most remarkable feature of the two examples in Fig. 5.28 is the chemical shift distribution in Me$_2$PP(CF$_3$)$_2$. The PMe$_2$ group in Me$_2$PP(CF$_3$)$_2$ has almost the same phosphorus resonance as in P$_2$Me$_4$, although the respective counter groups resonate at $\delta_P = -58.5$ ppm (Me$_2$P in P$_2$Me$_4$) and $\delta_P = 10.8$ (P(CF$_3$)$_2$ in Me$_2$PP(CF$_3$)$_2$).

This observation is unusual in diphosphanes, but rather normal in diphosphane monoxides, monosulfides, and monoselenides. Figure 5.29 gives the phosphorus

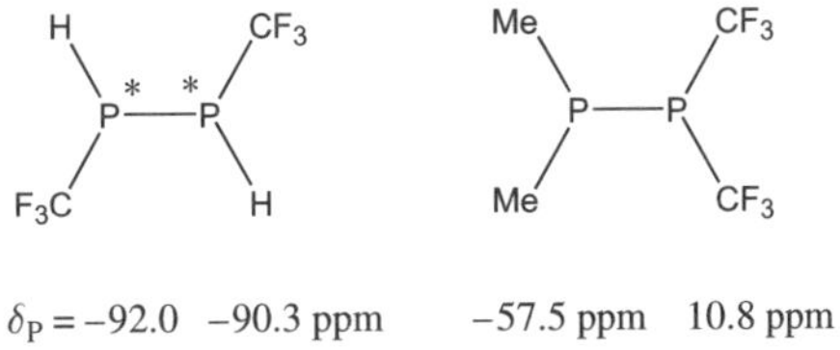

Fig. 5.28 ^{31}P chemical shift values for two diphosphanes with CF$_3$ substituents

$\delta_P = -92.0$ -90.3 ppm -57.5 ppm   10.8 ppm

$\delta_P = -58.7$ ppm 35.6 ppm; −57.5 ppm 21.9 ppm; 35.0 ppm; 33.9 ppm 14.6 ppm; 13.5 ppm

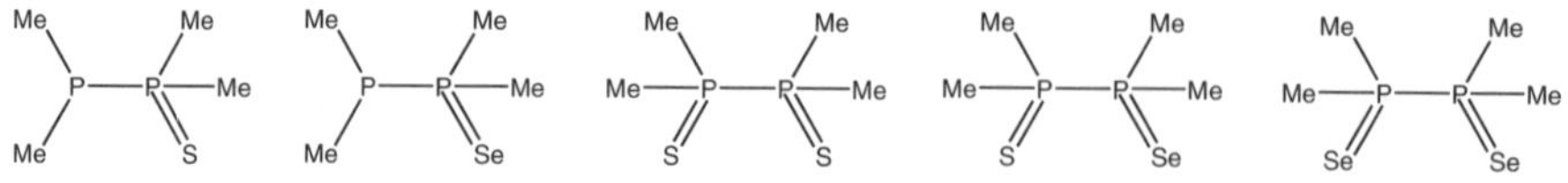

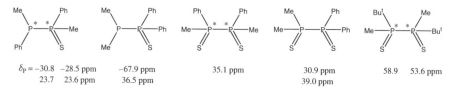

Fig. 5.29 Selected partially and fully oxidized tetramethyl diphosphanes and their ^{31}P chemical shift values

$\delta_P = -30.8$ −28.5 ppm 23.7 23.6 ppm; −67.9 ppm 36.5 ppm; 35.1 ppm; 30.9 ppm 39.0 ppm; 58.9 53.6 ppm

Fig. 5.30 Selected mixed partially and totally oxidized diphosphanes and their ^{31}P chemical shift values

chemical shifts of a selection of partially and totally oxidixed tetramethyl diphosphanes. In Fig. 5.30, these examples are complimented by a series of partially and totally oxidized mixed diphosphanes.

We note that $Me_2PP(S)Me_2$ features a chemical shift of $\delta_P = -58.7$ ppm virtually identical to that of P_2Me_4 at $\delta_P = -58.5$ ppm. The same holds true for the PMe_2 part of $Me_2PP(Se)Me_2$ at $\delta_P = -57.5$ ppm, and to a lesser extent for $Me_2PP(S)Ph_2$ ($\delta_P = -67.9$ ppm compared to $\delta_P = -64.2$ ppm), MePhPP(S)MePh ($\delta_P = -30.8$ ppm and −28.5 ppm compared to $\delta_P = -40.5$ ppm and −36.9 ppm), and $Ph_2PP(O)Ph_2$ ($\delta_P = -21.6$ ppm compared to $\delta_P = -14.7$ ppm).

We would have expected a downfield shift for both phosphorus atoms even if only one phosphorus atom is oxidized. The reason seems simple: oxidizing a phosphorus atom decreases the electron density on this atom. The oxidized phosphorus atom would then pull stronger on the electrons of the P-P single bond, causing a downfield shift in the resonance of the unoxidized phosphorus atom. In reality, we witness anything from a $\Delta\delta = 10$ ppm downfield to a $\Delta\delta = -7$ ppm upfield shift.

The NPPN compounds featuring an iminophosphorane and a phosphinoamido group depicted in Figs. 5.32–5.34 challenge our understanding of phosphorus chemical shifts. First, we need to determine which phosphorus atom gives rise to which

$\delta_P = -21.6$ ppm
39.2 ppm

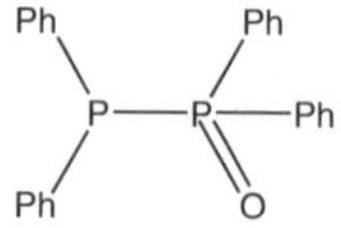

Fig. 5.31 Tetraphenyldiphosphane monoxide

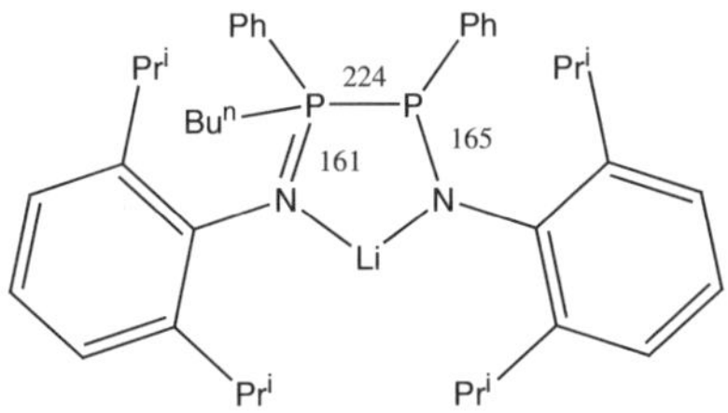

$\delta_P = 59.8$ ppm
26.9 ppm
$^1J_{PP} = 298$ Hz

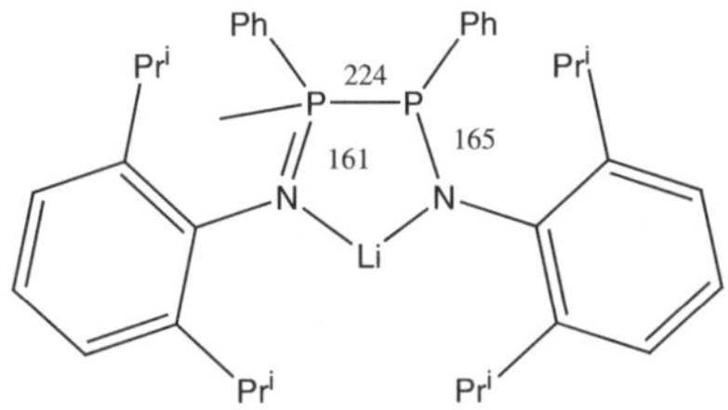

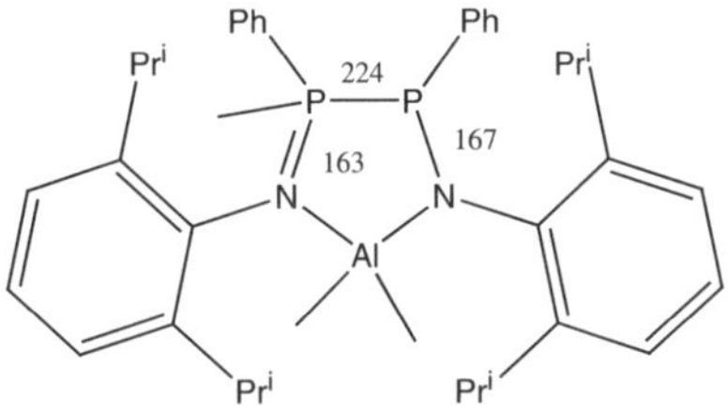

$\delta_P = 67.3$ ppm
19.0 ppm
$^1J_{PP} = 285$ Hz

45.9 ppm
40.2 ppm
264 Hz

Fig. 5.32 A series of metallated imino-amino diphosphanes

chemical shift. In Chap. 4 we have seen that iminophosphoranes have phosphorus resonances in the range of $\delta_P = -15$ to 10 ppm. Phosphinoamides normally resonate downfield from $\delta_P = 30$ ppm. This would place the chemical shifts of $\delta_P = 19.0$ ppm and $\delta_P = 26.9$ ppm in Fig. 5.32 with the iminophosphorane group, and the resonances of $\delta_P = 59.8$ ppm and $\delta_P = 67.3$ ppm with the phosphinoamido groups.

However, what is the explanation for the downfield shift of the phosphinoamido group and the upfield shift of the iminophosphorane group upon replacing

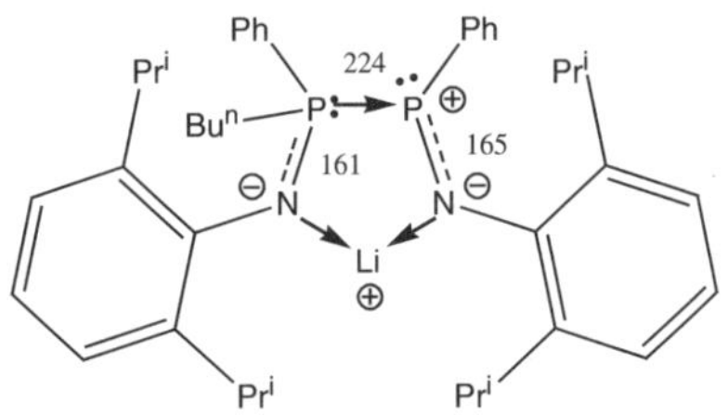

$\delta_P = 59.8$ ppm
26.9 ppm
$^1J_{PP} = 298$ Hz

Fig. 5.33 The lithiated imino-amino diphosphane depicted as a phosphino-phosphenium species

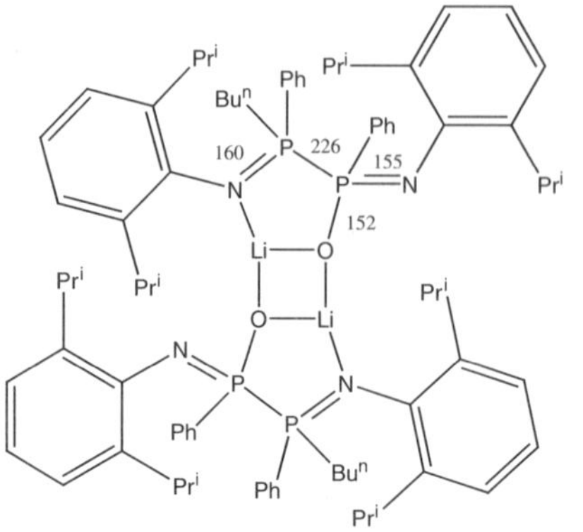

Fig. 5.34 The effect of
oxidation on the compound
shown in Fig. 5.33

$\delta_P = 5.3$ ppm
-10.2 ppm
$^1J_{PP} = 197$ Hz

the *n*-butyl substituent of the latter with a methyl group? The first observation of
interest is the short P—N bonds. The P—N bond lengths of 161 and 165 pm are
indicative of a substantial double bond character in these bonds. The P—P bond
length of 224 pm is close to the upper limit for a P—P bond, as in P_5Ph_5, a cyclic
pentaphosphane.

We are now in the predicament of explaining the expected upfield shift for
the iminophosphorane and the unexpected downfield shift for the adjacent phosphorus as we increase the electron density (methyl for *n*-butyl). Introducing the
methyl group increases the electron density on the iminophosphorane phosphorus
atom compared to the *n*-butyl derivative. This should result in a small increase of
electron density on the adjacent phosphorus atom by an inductive effect through
the P—P bond. It should also shift the P=N bond towards nitrogen. The result
is a downfield shift for the iminophosphorane and an upfield shift for the adjacent phosphorus atom, the exact opposite of what is observed. The way out is
to assume a weak π-interaction of the PALP along the P—P bond. This would
operate in the direction observed. However, the relatively long P—P bond length
speaks against it.

Another possible explanation is to assume a phosphino-phosphenium bond as
depicted in Fig. 5.33. This would actually mean that we experience a phosphinoamide
on the left side with a chemical shift in excess of $\delta_p = 30$ ppm involved in a donor bond
to a RN—PPh$^+$ species carrying a PALP. The latter is effectively an iminophosphane
carrying an extra lone pair. The positive charge rests on the donor phosphorus. The
chemical shifts of $\delta_p = 59.8$ ppm and $\delta_p = 67.3$ ppm might seem right, but the right
hand phosphorus should display chemical shifts significantly upfield to the observed
$\delta_p = 19.0$ ppm and $\delta_p = 26.9$ ppm. Introduction of the methyl group should cause an
upfield shift (alkyl increments, see page 2) in the left hand phosphorus (downfield is
observed), and an upfield shift (observed) in the right hand phosphorus.

The aluminium complex shown in Fig. 5.32 gives interesting insights into
the nature of the bonding. Here, both phosphorus atoms have almost the same
resonance, a clear indication that the electron densities have been nivillated. This

is best explained by hyperconjugation involving the PALP. A similar Zn-complex shows the same nivillation of the phosphorus resonances.

Oxidation of the PALP by oxygen creates two iminophosphorane entities with chemical shifts of $\delta_p = -10.2\,ppm$ and $\delta_p = 5.3\,ppm$, respectively. The P-P bond length increases to 226 ppm, indicative of a bond weakening, as there no longer is a PALP that can contribute to the bonding. The P=N bond lengths are 155 and 160 ppm, respectively, indicative of proper P=N double bonds. However, it should be remembered that the situation in the solid state is not necessarily preserved in solution.

Note: *Oxidation by sulfur rather than oxygen results in similar compounds with chemical shifts significantly downfield.*

5.3.2 Diphosphenes

Diphosphenes consist of two PR units fused together by a P=P double bond. They can be symmetrically or unsymmetrically substituted.

Table 5.10 shows the chemical shifts for P=P bonds. They can be found downfield of $\delta_p = 470\,ppm$, and are thus significantly downfield of P=C double bonds ($\delta_p = -100\,ppm$ to $\delta_p = 350\,ppm$). The factors responsible for the differences in the RN=PR$_3$, O=PR$_3$, S=PR$_3$ series operate also for phosphorus, and we will once again have to look for hyperconjugation effects to explain the trends in the phosphorus resonance accurately.

Note: *P=P double bonds have ^{31}P chemical shifts downfield of P=C double bonds.*

We have seen isolated P—P single and P=P double bonds. We ask ourselves what happens if the P=P double bond becomes part of a delocalized π-electron system, an aromatic ring.

A phosphole is essentially a cyclopentadiene where one of the CH units is substituted by a phosphorus atom. It results a compound of the formula C_4H_5P with two conjugated C=C double bonds. The chemical shift range is shown in Table 5.11, and can be found around $\delta_p = -50\,ppm$ in the upfield section of the spectrum, in close proximity to secondary phosphanes.

Deprotonation creates a phospholide analogous to a cyclopentadienide, a five-membered aromatic ring system. Whether C_4H_4P-rings are indeed aromatic has been discussed at length, and can be answered in the affirmative. Figure 5.35 shows the respective resonance structures.

Note: *Deprotonation results in a $\Delta\delta = 110-140\,ppm$ downfield shift as the phosphorus atom becomes part of a P=C double bond system. The increase in electron density should result in an upfield shift.*

Note: *Consecutive substitution of a CH group by phosphorus in the $C_5H_5^-$ system results in consecutive downfield shifts.*

Table 5.10 The ^{31}P chemical shift values of selected P=P bonds

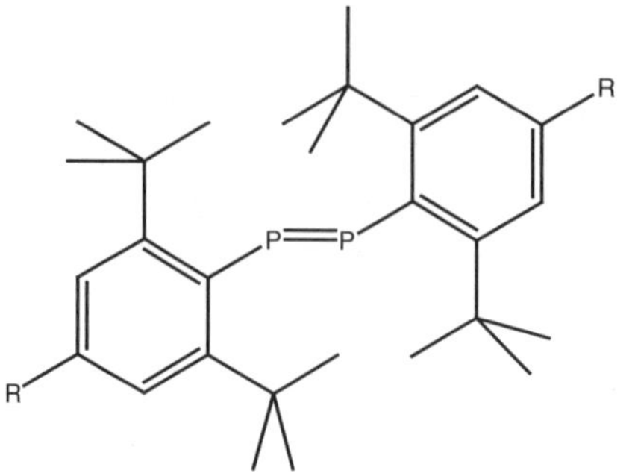

R	δ_P [ppm]
H	488.7
Bu^t	490.0
OMe	495.9
NMe$_2$	500.6

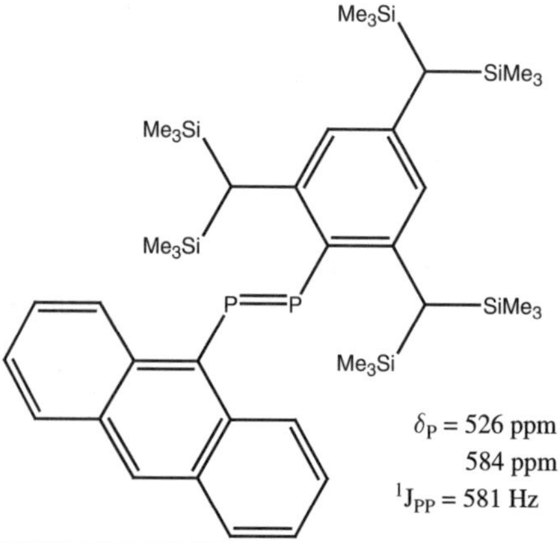

δ_P = 526 ppm
584 ppm
$^1J_{PP}$ = 581 Hz

Table 5.11 ^{31}P chemical shift values for a series of phospholides

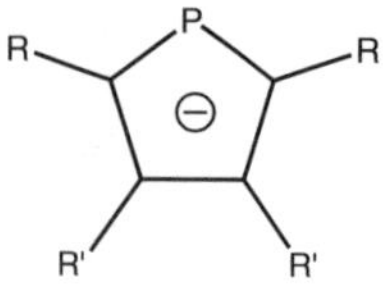

R	R'	δ_P	δ_P H-P
H	H	76.6	−49.2
H	Me	58.9	−59.4
Ph	H	78.7	−54.2
Ph	Ph	98.8	−40.9

This is depicted in Fig. 5.36, where we can see the stepwise downfield shifts each time another phosphorus atom is introduced into the ring.

Note: *P$_5^-$ (and the other phospholes) resonates upfield (δ_P = 470 ppm) from the diphosphene range (δ_P = 480–600 ppm).*

An example for the downfield shift caused by phosphorus inclusion into the phosphole ring is shown in Fig. 5.37. The substituents on carbon are in both

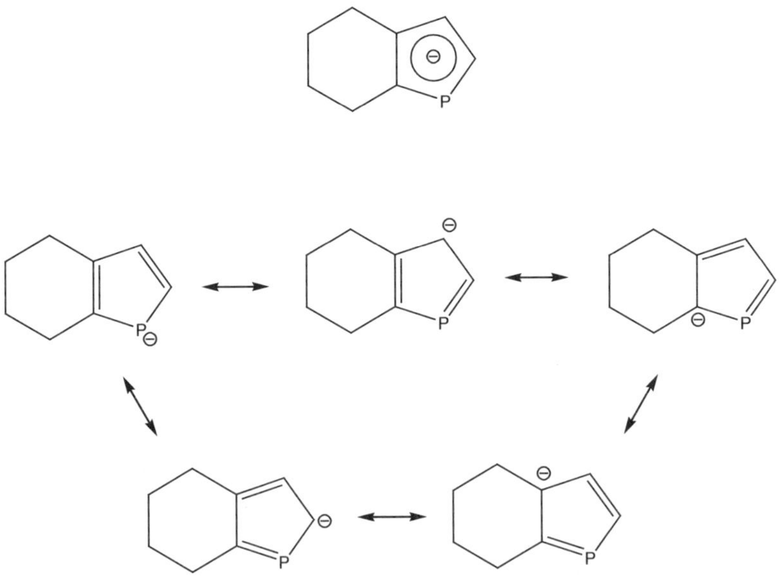

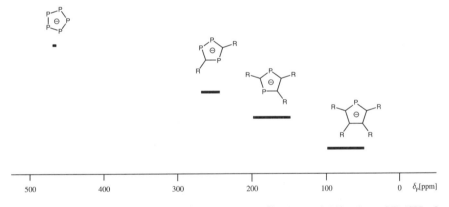

Fig. 5.35 Resonance structures for an anellated phospholide

Fig. 5.36 The influence of the phosphorus content on the ^{31}P chemical shift values of $[P_n(CH)_{5-n}]^-$

cases identical. The two phosphorus atoms in $(CBu^t)_3P_2^-$ are in identical chemical environments, resulting in a solitary singlet at $\delta_p = 187.6$ ppm. Inclusion of a third phosphorus atom necessitates a direct P-P bond and two different chemical environments. The two phosphorus resonances are very similar at $\delta_p = 245.5$ ppm (d) and $\delta_p = 252.5$ ppm (t), about $\Delta\delta = 60$ ppm downfield from $(CBu^t)_3P_2^-$.

Fig. 5.37 Comparison between a 1,3,4-triphosphaphospholide and a 1,3-diphosphaphospholide

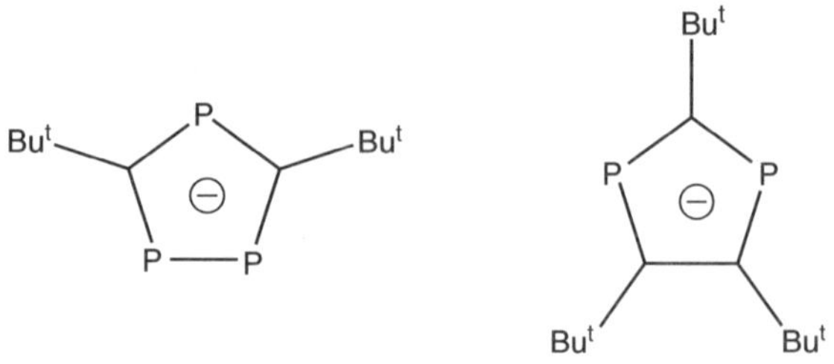

$\delta_P = 245.5$ ppm, d
252.5 ppm, t
$^2J_{PP} = 47$ Hz

187.6 ppm

Bibliography

Agbossou F, Carpentier J-F, Hapiot F, Suisse I, Mortreux A, Coord Chem Rev 178–80 (1998) 1615.

Agbossou-Niedercorn F, Suisse I, Coord Chem Rev 242 (2003) 145.

Albrand J P, Robert J B, Goldwhite H, Tetrahedron Lett (1976) 949.

Anagho L E, Bickley J F, Steiner A, Stahl L, Angew Chem Int Ed 44 (2005) 3271.

Avens L R, Cribbs L V, Mills J L, Inorg Chem 28 (1989) 205.

Bansal R K, Heinicke J, Chem Rev 101 (2001) 3549.

Bartsch R, Nixon J F, J Organomet Chem 415 (1991) C15.

Baudler M, Akpapoglou S, Ouzounis D, Wasgestian F, Meinigke B, Budzikiewicz H, Münster H, Angew Chem Int Ed 27 (1988) 280.

Baudler M, Zarkadas A, Chem Ber 105 (1972) 3844.

Bettermann G, Schmutzler R, Pohl S, Thewalt U, Polyhedron 6 (1987) 1823.

Bhattarcharyya P, Slawin A M Z, Smith M B, Williams D J, Woollins J D, J Chem Soc, Dalton Trans (1996) 3647.

Braunstein P, Frison C, Morise X, Adams R D, J Chem Soc, Dalton Trans (2000) 2205.

Breen T L, Stephan D W, J Am Chem Soc 117 (1995) 11914.

Charrier C, Bonnard H, de Lauzon G, Matthey F, J Am Chem Soc 105 (1983) 6871.

Chivers T, Copsey M C, Fedorchuk C, Pavez M, Stubbs M, Organometallics 24 (2005) 1919.

Chivers T, Copsey M C, Pavez M, Chem Comm (2004) 2818.

Copsey M C, Chivers T, Dalton Trans (2006) 4114.

Cowley A H, Hall S W, Polyhedron 8 (1989) 849.

Daly J J, J Chem Soc (1964) 6147.

Fei Z, Scopelliti R, Dyson P J, Eur J Inorg Chem (2004) 530.

Ferguson G, Myers M, Spalding T R, Acta Cryst C46 (1990) 122.

Fernandez A, Reyes C, Lee T Y, Prock A, Giering W P, Haar C M, Nolan S P, J Chem Soc, Perkin Trans II (2000), 1349.

Fluck E, Binder H, Inorg Nucl Chem Lett 3 (1967) 307.

Fluck E, Lorenz J, Z Naturforsch 22b (1967) 1095.

Grim S O, McFarlane W, Nature 208 (1965) 995.

Gruber M, Jones P G, Schmutzler R, Chem Ber 123 (1990) 1313.

Gruber M, Jones P G, Schmutzler R, Phosphorus 80 (1993) 195.

Gruber M, Schmutzler R, Phosphorus 80 (1993) 181.Gruber M, Schmutzler R, Schomburg D, Phosphorus 80 (1993) 205.

Heinicke J, Gupta N, Surana A, Peulecke N, Witt B, Steinhauser K, Bansal R K, Jones P G, Tetrahedron 57 (2001) 9963.

Ito S, Freytag M, Yoshifuji M, Dalton Trans (2006) 710.

Ito S, Miyake H, Sugiyama H, Yoshifuji M, Heteroatom Chem 16 (2005) 357.

Ito S, Miyake H, Sugiyama H, Yoshifuji M, Tetrahedron Lett 45 (2004) 7019.

Ito S, Miyake H, Yoshifuji M, Höltzl T, Veszprémi T, Chem Eur J 11 (2005) 5960.

Ito S, Sugiyama H, Yoshifuji M, Angew Chem Int Ed 39 (2000) 2781.

Kawasaki S, Nakamura A, Toyota K, Yoshifuji M, Bull Soc Chem Jpn 78 (2005) 1110.

Klebach T C, Lourens R, Bickelhaupt F, J Am Chem Soc 100 (1978) 4886.

Kühl O, Coord Chem Rev 249 (2005) 693.

Kühl O, Coord Chem Rev 250 (2006) 2867.

Kühl O, Blaurock S, Hey-Hawkins E, Z Anorg Allg Chem 625 (1999) 1517.

Kühl O, Blaurock S, Sieler J, Hey-Hawkins E, Polyhedron 20 (2001) 111.

Kühl O, Junk P C, Hey-Hawkins E, Z Anorg Allg Chem 626 (2000) 1591.

Kühl O, Koch T, Somoza F B Jr, Junk P C, Hey-Hawkins E, Plat D, Eisen M S, J Organomet Chem 604 (2000) 116.

Ly T Q, Slawin A M Z, Woollins J D, Angew Chem 110 (1998) 2605.

Ly T Q, Slawin A M Z, Woollins J D, Polyhedron 18 (1999) 1761.

Maier L, Helv Chim Acta 49 (1966) 1718.

Matthey F, Coord Chem Rev 137 (1994) 1.

McFarlane H C E, McFarlane W, Nash J A, J Chem Soc, Dalton Trans (1980) 240.

Murakami F, Sasaki S, Yoshifuji M, J Am Chem Soc 127 (2005) 8926.

Nishida K, Liang H, Ito S, Yoshifuji M, J Organomet Chem 690 (2005) 4809.

Quin L D, Orton W L, J Chem Soc, Chem Comm (1979) 401.

Rodriguez i Zubiri M, Milton H L, Slawin A M Z, Woollins J D, Polyhedron 23 (2004) 865.

Sasamori T, Mieda E, Nagahora N, Takeda N, Takagi N, Nagase S, Tokitoh N, Chem Lett 34 (2005) 166.

Sasamori T, Takeda N, Tokitoh N, J Phys Org Chem 16(2003) 450.

Sasamori T, Tsurusaki A, Nagahora N, Matsuda K, Kanemitsu Y, Watanabe Y, Furukawa Y, Tokitoh N, Chem Lett 35 (2006) 1382.

Schöller W W, Stämmler V, Inorg Chem 23 (1984) 3369.

Slawin A M Z, Wainwright M, Woollins J D, J Chem Soc, Dalton Trans (2001) 2724.

Sugiyama H, Ito S, Yoshifuji M, Chem Eur J 10 (2004) 2700.

Tolman C A, Chem Rev 77 (1977) 313.

Toyota K, Kawasaki S, Yoshifuji M, J Org Chem 69 (2004) 5065.

Vogt R, Jones P G, Kolbe A, Schmutzler R, Chem Ber 124 (1991) 2705.

Chapter 6
Main Group Compounds

We would expect that phosphanes can utilize their electron lone pair to bond to Lewis acids (both from transition metals and from main group elements). However, they can also act as Lewis acids. The best known example is probably PF_5, but similar molecules, like $POCl_3$ and PCl_5 are known. PCl_5 is present as $[PCl_4][PCl_6]$ in the solid state, the result of PCl_5 acting as a Lewis acid toward itself, creating a PCl_6^- anion by abstraction of Cl^- and leaving a PCl_4^+ cation behind.

At left, the ^{31}P-NMR references of phosphorus chlorides are depicted in order of decreasing coordination number of phosphorus. However, the coordination number is not expected to be the only ordering principle, as PCl_4^- would fall outside the range of PCl_4^+ and the phosphonium salts are seen to resonate significantly upfield from PCl_4^+.

The chapter is divided into Lewis basic behavior and Lewis acidic behavior, respectively. The concept overlaps somewhat at the end of Sect. 6.1, as we examine examples where both the Lewis base and the Lewis acid are phosphorus containing species.

Utilization of PF_5 as a fluoride abstracting Lewis acid is also mentioned in Chap. 7, where the fluoride is abstracted from a fluorophosphane bonded to a transition metal.

6.1 As Lewis Base

Phosphanes are characterized, among other things, by their electron lone pair. This electron lone pair can be expected to be utilized in a σ-donor interaction toward a Lewis acid, making the phosphane a Lewis base. In fact, that is the reason for the popularity of phosphanes in transition metal chemistry. Of course, the Lewis basicity not only makes them good ligands, but lets phosphanes develop a rich and diverse main group chemistry as well.

The most obvious choices for a Lewis acid to exploit the Lewis basicity of phosphanes are group 13 elements with their intrinsic electron deficiency. Looking at BH_3 as the Lewis acid component, we can easily discern the trends in the Lewis basicity of phosphanes. In the top part of Table 6.1, the phosphanes experience a pronounced coordination chemical shift of $\Delta\delta = 60-135\,ppm$ from a well-shielded

O. Kühl, *Phosphorus-31 NMR Spectroscopy*,
© Springer-Verlag Berlin Heidelberg 2008

Fig. 6.1 Chemical shift
values for chlorophosphorus
complexes

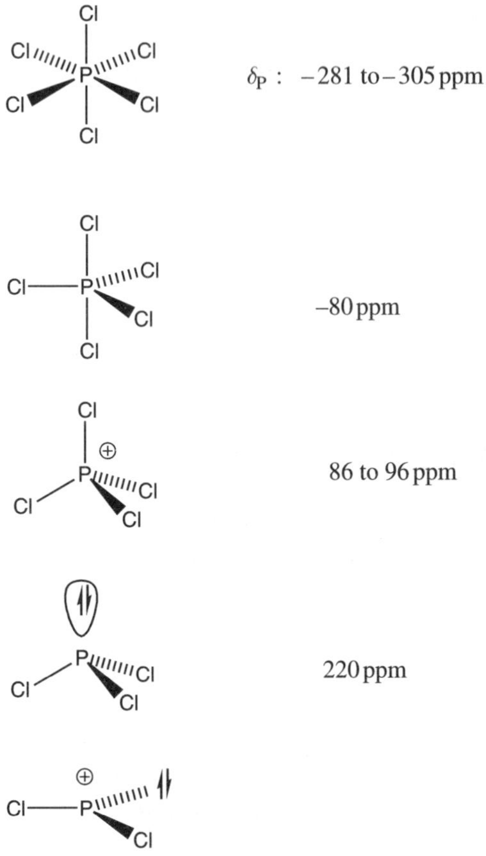

δ_P : −281 to−305 ppm

−80 ppm

86 to 96 ppm

220 ppm

Table 6.1 ^{31}P-NMR resonances and coordination chemical shift values for phosphino boranes

Compound	δ_p phosphane [ppm]	δ_p complex [ppm]	$\Delta\delta$ [ppm]
PH$_3$BH$_3$	−246	−113	133
MePH$_2$BH$_3$	−163.5	−68.5	95
PhPH$_2$BH$_3$	−123.5	−49.3	74.2
Me$_2$PHBH$_3$	−98.5	−30.8	67.7
Me$_3$PBH$_3$	−62.8	−1.8	61
PhMe$_2$PBH$_3$	−46	49	95
(MeO)PF$_2$BH$_3$	111.8	108.5	−3.3
PF$_3$BH$_3$	105	107	2
(Me$_2$N)$_3$PBH$_3$	122.5	102.5	−20
(MeO)$_2$PFBH$_3$	131.6	118.7	−12.9
(MeO)$_3$PBH$_3$	140	118	−22
(Me$_2$N)PF$_2$BH$_3$	143	130	−13
(Me$_2$N)$_2$PFBH$_3$	153	134	−19
(CF$_3$)PF$_2$BH$_3$	158.1	148.5	−9.6

Table 6.2 ^{31}P-NMR resonances and coordination chemical shift values for phosphane chloro-gallane adducts

R	δ_P (R$_2$ClPGaCl$_3$) [ppm]	δ_P (R$_2$ClP) [ppm]	$\Delta\delta$ [ppm]
Ph	41	81.5	−40.5
Me	57	92	−35
Et	79	119	−40
Pr^i	91	–	–
Bu^t	101	145	−44

resonance of $\delta_P = -46$ to -246 ppm as free ligands. In stark contrast, the phosphanes in the lower part of the table experience a moderate upfield coordination chemical shift of $\Delta\delta = -10$ to -22 ppm, with PF$_3$BH$_3$ and (MeO)PF$_2$BH$_3$ somewhat in between with $\Delta\delta = 2$ and -3.3 ppm, respectively.

We would expect a considerable downfield shift upon coordination of the phosphane, and we are therefore not surprised to observe it in the ensuing adducts. However, why do we observe an upfield shift upon coordination to the borane with the phosphanes in the lower part of the table? The difference must lie in the behavior of the substituents on phosphorus, as this is the one parameter that changes as we look down the list. In the top part, the substituents are H, methyl, and phenyl, whereas in the lower part, the substituents are fluoride, amide, and methoxide. The latter three (F, NMe$_2$ and MeO) are capable of a π-bonding interaction toward phosphorus that increases as the electron density on phosphorus diminishes upon coordination. Since the ^{31}P-NMR chemical shifts are more sensitive toward π-interactions than σ-interactions, the net result can very well be an upfield shift upon coordination of the phosphane, if substituents capable of "π-backbonding" are present on phosphorus.

This argument is confirmed by a series of monochloro phosphane gallium(III) chloride adducts. The chemical shift values for the free ligand and the Ga(III) adduct change in accord with the alkyl or aryl substituent on phosphorus, but the coordination chemical shift stays in a very narrow range: $\Delta\delta = -35$ to -44 ppm upfield from the free ligand, indicative of a π-bonding contribution from the P—Cl substituent.

This Lewis base behaviour is not limited to group 13 complexes, but can be observed with all main group Lewis acids. A particularly interesting example is the intramolecular Lewis basicity toward another phosphorus group in bisphosphino ureas and thioureas. In Fig. 6.2, the PPh$_2$ group acts as a Lewis base toward the PF$_2$ group. The $^1J_{PP}$ coupling constant of 110 Hz can have its origin in a somewhat weak interaction that does not quite amount to a full single bond, or "through space" interactions of the two lone pairs. However, addition of PF$_5$ results in fluoride abstraction (PF$_5$ acts as a Lewis acid, see Sect. 6.2), and a proper P—P bond is formed ($^1J_{PP} = 304$ Hz). Interestingly, fluoride abstraction and formation of the cation results in an upfield shift for **both** phosphorus atoms, while the formal coordination number changes on **one** phosphorus atom only. The likely explanation is again a π-bonding interaction from fluorine and/or nitrogen. Hyperconjugation predicts donation from a non-bonding fluorine or nitrogen orbital into an antibonding P—P orbital, thus increasing the electron density on **both** phosphorus atoms.

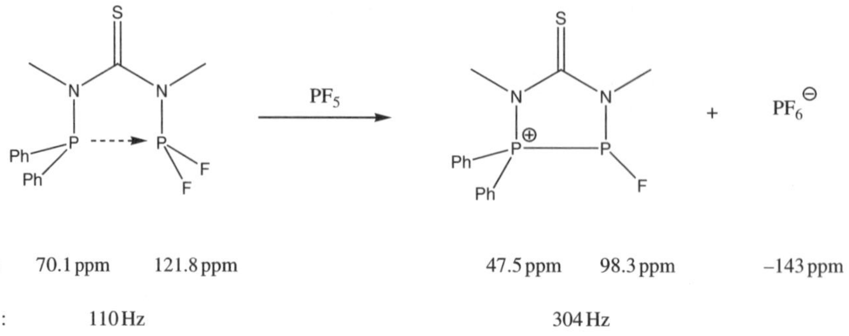

δ_P: 70.1 ppm   121.8 ppm   47.5 ppm   98.3 ppm   −143 ppm

J_{PP}:   110 Hz   304 Hz

Fig. 6.2 Fluoride abstraction by PF$_5$ to form a phosphino phosphenium cation

The same intramolecular Lewis base – Lewis acid interaction can be observed when a chlorophosphane is used instead of a fluorophosphane. However, the chloride is less strongly bonded than fluoride, resulting in the displacement of chloride by the phosphane without the use of an auxiliary Lewis acid. The chemical shift of the tricoordinate phosphorus atom is sensitive to the steric bulk of its carbon substituent. Evidently, sterically demanding substituents like *tert*-butyl hinder the π-bonding interaction from nitrogen, resulting in the observed downfield shift.

Phosphanes react with alkyl and aryl halides to form phosphonium salts. Their phosphorus chemical shifts are in a narrow range at $\delta_p = 20$–60 ppm (Ph$_4$P$^+$: $\delta_p = 20$ ppm; Bu$_4^t$P$^+$ $\delta_p = 58$ ppm).

The lone pair on phosphorus is at the centre of its main group chemistry. The phosphorus atom can act as a Lewis acid when it is cationic or in an oxidation state other than +III (most likely +II or +I), but can also react with virtually any Lewis acid, including itself.

Table 6.3 Intramolecular Lewis base – Lewis acid behavior of phosphanes

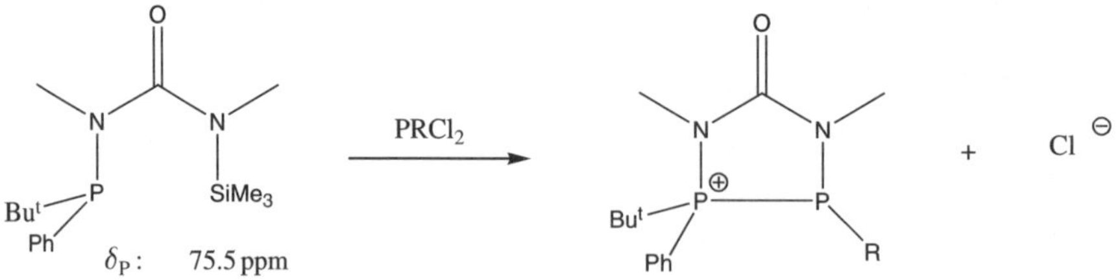

R	δ_p (P$^+$) [ppm]	δ_p (P) [ppm]	$^1J_{PP}$ [Hz]
Me	61.0	−12.0	310
Et	61.5	−5.3	303
Pr^i	59.6	12.0	304
Bu^t	55.4	36.2	302
Ph	59.8	1.7	278
CHCl$_2$	52.9	−11.4	304
CH$_2$SiMe$_3$	65.6	8.1	333

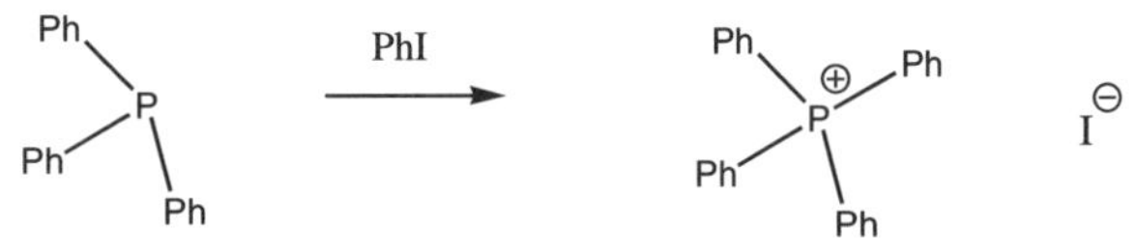

Fig. 6.3 Formation of phosphonium cations

6.2 As Lewis Acid P(I), P(III), and P(V)

Arguably the best known phosphorus containing Lewis acid is PF_5. It is often used to abstract a fluoride ion from another molecule, thus forming PF_6^-, a popular non-coordinating anion. Its ^{31}P-NMR resonance is observed at around $\delta_p = -144$ ppm.

If this other molecule is a fluorophosphane, a phosphenium cation is generated that is also a Lewis acid, but by necessity weaker than PF_5. Halide abstraction from halophosphanes is a very popular method to obtain phosphenium Lewis acids.

Reaction of PF_4R (R=Me, Ph, F) with a carbene results in the six-coordinate phosphorus species PF_4R(carbene). The phosphorus compound acts as a Lewis acid toward the Lewis basic carbene, and the phosphorus resonance is shifted upfield by about $\Delta\delta = -100$ ppm. The fine structure of this upfield shift is of considerable interest. The electronegativity of the substituent R increases in the order Me < Ph < F in accord with an upfield shift in the phosphorus resonance. Therefore, we again witness a case where the loss in electron density through the σ-backbone is partially compensated by an increased π-donor interaction, in the present case from the fluoride substituents on phosphorus. As the influence of the π-bonding interaction on the phosphorus chemical shift is larger than that of the σ-bonding interaction, we witness a net upfield shift.

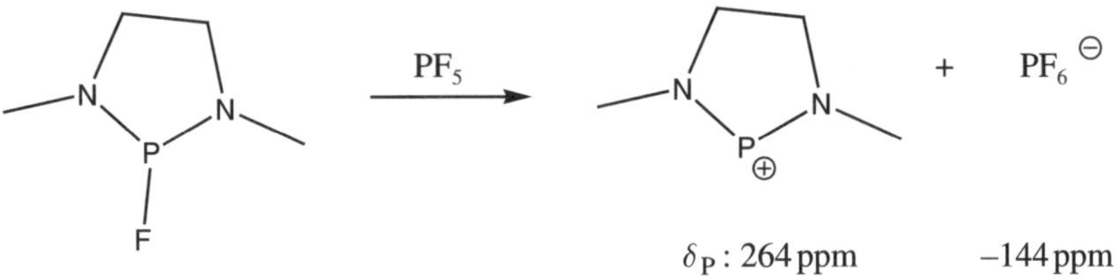

δ_P: 264 ppm -144 ppm

Fig. 6.4 Fluoride abstraction by the Lewis acid PF_5

Table 6.4 Fluorophosphorane carbene adducts

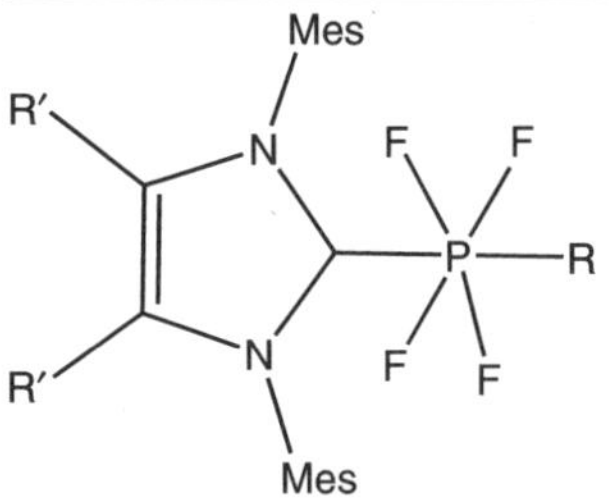

	R	R′	δ_p [ppm]
	Me	H	−127.05
	Ph	H	−141.06
	F	H	−148.40
	F	Cl	−151.79

The argument is strengthened by comparison of the two carbenes. Introduction of chlorine atoms in 4,5-position lowers the nucleophilicity of the carbene. In turn, the electron density on phosphorus decreases, and we would expect a downfield shift of the resonance. However, we see a small upfield shift instead due to increased π-donation from the fluorine atoms. This trend is corroborated by the shortening of the P-F bonds by $\approx 2\,\mathrm{pm}$ going from R = Ph to R' = Cl.

The same trend can be seen in the series of five-coordinate phosphorus compounds depicted in Table 6.5. Decrease of σ-donation along the series $Me_3SiCH_2 > Me > 2,5\text{-}Me_2C_6H_3 > Ph$ is accompanied by an increase of hyperconjugation from equatorial ligands (amine, F, pyrrole, Ph), resulting in an overall upfield shift.

Ring closure to the cationic species yields the expected downfield shift due to the introduction of a positive charge. The downfield shift is very moderate, because of the compensation from the amine functionality and the substituent R. The trend established by the substituent R remains essentially unchanged.

Note: *In the cationic compound, the pyrrole group is in an axial position, and thus cannot contribute to a π-bonding interaction. In consequence, its phosphorus resonance is at the downfield end of the series, whereas it is at the upfield end of the neutral species.*

The hypothetical phosphenium cation, PPh_2^+, a P(III) species, has recently attracted great interest. It acts as Lewis acid toward a range of Lewis bases, amongst which the phosphanes and carbenes are possibly the best known. The phosphorus resonance of the PPh_2 part of the ensuing Lewis acid – Lewis base adduct is shifted upfield in accord with the nucleophilicity of the Lewis base employed. The Lewis base part, however, does not follow such a clear trend. In particular, PMe_3 shows the same chemical shift value as PPh_3, an effect that is consistently observed in similar adducts.

Table 6.5 ^{31}P-NMR chemical shifts for λ^5-fluorophosphoranes

	R'	R	δ_P [ppm]
	F	Me_3SiCH_2	−33.0
	F	Me	−35.6
	F	$2,5\text{-}Me_2C_6H_3$	−44.5
	Ph	Ph	−51.9
	F	Ph	−58.0
	C_4H_4N	Ph	−66.4

Table 6.6 ^{31}P-NMR chemical shifts for intramolecularly Lewis base stabilized λ^5-phosphenium cations

	R'	R	δ_P [ppm]
	F	Me_3SiCH_2	−13.3
	F	Me	−17.0
	F	Ph	−26.9
	C_4H_4N	Ph	−9.8

Table 6.7 ^{31}P-NMR chemical shift values for phosphino phosphenium cations

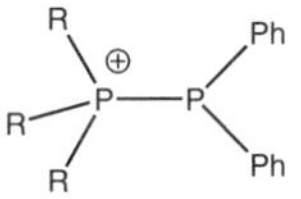

		R	Ph	Cy	Me	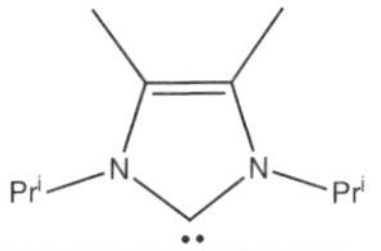
		δ_p (R$_3$P)	15	25	15	–
		δ_p (PPh$_2$)	–10	–21	–23	–27
		$^1J_{PP}$ [Hz]	350	361	289	–

Note: *In the amine substituted phosphino-phosphenium cations, the four-coordinate phosphorus atom carrying the cationic charge resonates upfield from the tricoordinate phosphorus atom, while the reverse is the case in the phenyl substituted phosphino-phosphenium cations. The effect of the π-bonding interaction of the amino groups is clearly visible.*

Of considerably greater interest from a spectroscopic point of view is the series of acyclic phosphenium cations shown in Fig. 6.6. Here, the phosphorus resonance is controlled by substituent effects, and ranges from δ_p=264 to δ_p=513 ppm. Diversion or hindrance of π-donation ability of the nitrogen substituents causes a noticeable downfield shift in the phosphorus resonance. Substitution of a dimethylamine functionality by a chloride causes a downfield shift of $\Delta\delta$=61 ppm, whereas the substitution of the peripheral methyl groups on a dimethylamine functionality by silyl groups causes a considerably larger downfield shift of $\Delta\delta$=90.3 ppm. The reason is that silicon is a far better π-acceptor toward the nitrogen atom than phosphorus. This is corroborated by the structures of E(SiMe$_3$)$_3$ (E=N, P), the amine is planar while the phosphine is not. As a result, the π-interaction is diverted from phosphorus to silicon, with a subsequent downfield shift in the phosphorus resonance. The effect is clearly additive, as successive substitution of the remaining two methyl groups by silyl functionalities results in an addidtional downfield shift of $\Delta\delta$=96 ppm and $\Delta\delta$=186.3 ppm, respectively.

One would expect that substitution of one dimethylamine functionality by a *tert*-butyl group would result in a very moderate downfield shift similar to the one caused by chlorine substitution. However, the downfield shift is a staggering $\Delta\delta$=249.2 ppm, and thus the greatest observed in the series. What is at first surprising becomes clear upon closer inspection. In order for effective π-donor bonding to

δ_P: 50 ppm   118 ppm   53 ppm   94 ppm

Fig. 6.5 ^{31}P-NMR chemical shifts for some amino-substituted phosphino phosphenium cations

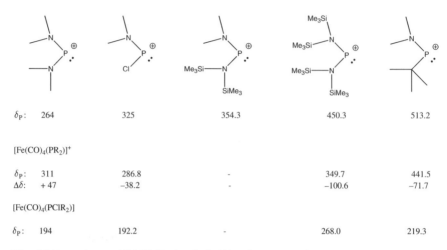

δ_P: 264   325   354.3   450.3   513.2

$[Fe(CO)_4(PR_2)]^+$

δ_P: 311   286.8   -   349.7   441.5
$\Delta\delta$:   + 47   −38.2   -   −100.6   −71.7

$[Fe(CO)_4(PClR_2)]$

δ_P: 194   192.2   -   268.0   219.3

Fig. 6.6 Dependance of ^{31}P-NMR chemical shift values upon substitution in a series of amino phosphenium cations

occur between phosphorus and nitrogen, the methyl groups on nitrogen have to be in plane with the trigonal planar phosphorus atom, and would then collide with the methyl groups of the *tert*-butyl group. Steric crowding thus prevents π-donor bonding and causes the dramatic downfield shift. A description of the effect using the concept of hyperconjugation would discuss the chemical shift differences in terms of angle dependency. Of course, the largest upfield shift would be observed in the event of coplanarity (180°) in agreement with the π-donor concept.

Comparison of the free phosphenium cations with their $Fe(CO)_4$ adducts is again very instructive. All phosphenium cations experience an upfield shift upon coordination to the $Fe(CO)_4$ fragments (despite the strongly π-accepting carbonyl groups), with the exception of $(NMe_2)_2P^{+\cdot}$ which experiences a downfield shift of $\Delta\delta = 47$ ppm like an ordinary phosphane. Of course, with the π-bonding interaction of two dimethylamine functionalities already in place, backbonding from the metal is no longer substantial. M-P backbonding can be explained by hyperconjugation. However, the orbitals involved on phosphorus would be the same as those in the P-N interactions, and thus already engaged.

A similar system to the P(V) phosphoranes we have experienced in Tables 6.5 and 6.6 is also available for phosphorus (III) and presented in Table 6.8. The trends are absolutely analogous, with the exception of the different influences of equatorial and axial ligands, as such a distinction does not apply in λ^3-phosphanes. The phosphorus resonances are shifted downfield by $\Delta\delta = 150$–200 ppm in accord with the difference in coordination numbers on phosphorus.

It can again be seen that the σ-withdrawing effect Me $<$ Ph $<$ CCl$_3$ $<$ CF$_3$ is overcompensated by the additional π-bonding interaction from the nitrogen substituents, and that the same modulate the expected downfield shift upon introduction of a positive charge to a mere $\Delta\delta = 10$–30 ppm.

Table 6.8 Coordination chemical shifts upon formation of cyclo λ^3-amino phosphenium cations

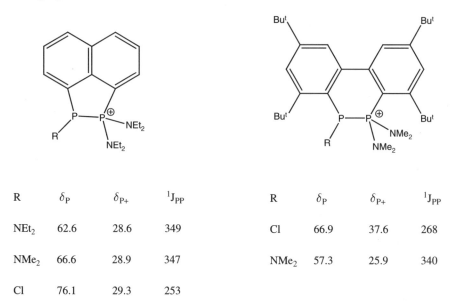

Y	δ_p [ppm]	δ_p [ppm]	$\Delta\delta$ [ppm]
Me	149.8	179.7	29.9
Ph	143.8	165.0	21.2
CC^{l3}	118.5	130.2	11.7
CF^3	98.6	119.3	20.7

We have already seen in the case of phosphino ureas that the interaction between a Lewis basic phosphane and a Lewis acidic phosphenium cation is not limited to intermolecular examples, but also occurs intramolecularly between neighboring phosphorus atoms in the same molecule. Two interesting examples are presented in Fig. 6.7. The phosphorus resonances are largely independent of the ring size (five- or six-membered), although the resonance of the neutral phosphorus atoms in the six-membered ring seem to be $\Delta\delta=-10$ ppm upfield from those in the five-membered ring. The $^1J_{PP}$ coupling constant predictably increases by 70–80 Hz upon changing the chloride substituent to an amino function.

R	δ_P	δ_{P+}	$^1J_{PP}$
NEt$_2$	62.6	28.6	349
NMe$_2$	66.6	28.9	347
Cl	76.1	29.3	253

R	δ_P	δ_{P+}	$^1J_{PP}$
Cl	66.9	37.6	268
NMe$_2$	57.3	25.9	340

Fig. 6.7 Cyclic phosphino phosphenium cations with aromatic backbones

Table 6.9 Phosphino phosphenium cations with P(III) and P(I) cores

Compound	δ_p (PPh$_3$) [ppm]	δ_p (P) [ppm]	$^1J_{PP}$ [Hz]
Ph$_2$P-PPh$_2$	–	–14	–
[Ph$_3$P-PPh$_2$]$^+$	15	–10	343
[Ph$_3$P-P-PPh$_3$]$^+$	30	–174	502
[Ph$_3$P-PH-PPh$_3$]$^{2+}$	23	–120	286

Going from a phosphenium (III) to a phosphenium (I) cation does not change the general concepts in the Lewis acidity of the cations, but causes a substantial upfield shift of $\Delta\delta = -150$ to -250 ppm as one moves from P(III) to P(I). This is not surprising, as the change in oxidation state in this case is equivalent to an additional electron lone pair on P(I).

Lewis base stabilization by phosphanes can lead to cyclic or acyclic species, with the cyclic compounds resonating some $\Delta\delta = -30$ to -60 ppm upfield from the acyclic compounds. With similar substituents on the P(I) atom, the magnitude of the upfield shift seemingly depends on the P(III)-P(I)-P(III) bond angle. Increasing the bond angle shifts the phosphorus resonance downfield as P(I)-P(III) backbonding becomes more feasible.

Protonation of the phosphenium cation occurs at the central P(I) phosphorus atom that has acquired a partial negative charge due to the σ-donor interaction with the flanking P(III) atoms. The resonance is duly shifted downfield by $\Delta\delta = 54$ ppm, and the $^1J_{PP}$ coupling constant almost halved from 502 Hz to 286 Hz, respectively. Despite the introduction of a second positive charge (protonation of a cation), the resonance of the P(III) centre is actually shielded by $\Delta\delta = -7$ ppm as one goes from [Ph$_3$P-P-PPh$_3$]$^+$ to [Ph$_3$P-PH-PPh$_3$]$^{2+}$.

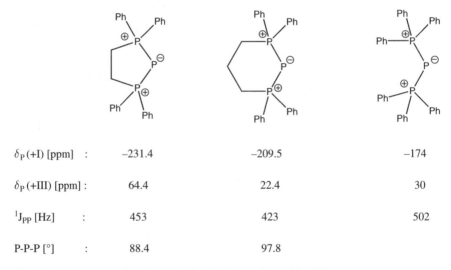

δ_P (+I) [ppm] :	–231.4	–209.5	–174
δ_P (+III) [ppm] :	64.4	22.4	30
$^1J_{PP}$ [Hz] :	453	423	502
P-P-P [°] :	88.4	97.8	

Fig. 6.8 Cyclic and acyclic phosphino phosphenium cations with a P(I) core

Bibliography

Atwood J L, Robinson K D, Bennett F R, Elms F M, Koutsantonis G A, Raston C L, Young D J, Inorg Chem 31 (1992) 2673.

Bennett F R, Elms F M, Gardiner M G, Koutsantonis G A, Raston C L, Roberts N K, Organo metallics 11 (1992) 1457.

Burford N, Cameron T S, LeBlanc D J, Losier P, Sereda S, Wu, G Organometallics 16 (1997) 4712.

Burford N, Dyker C A, Phillips A D, Spinney H A, Decken A, McDonald R, Ragogna P J, Rheingold A L, Inorg Chem 43 (2004) 7502.

Burford N, Herbert D E, Ragogna P J, McDonald R, Ferguson M J, J Am Chem Soc 126 (2004) 17067.

Burford N, Losier P, Phillips A D, Ragogna P J, Cameron T S, Inorg Chem 42 (2003) 1087.

Burford N, Phillips A D, Spinney H A, Lumsden M, Werner-Zwanziger U, Ferguson M J, McDonald R, J Am Chem Soc 127 (2005) 3921.

Burford N, Ragogna P J, J Chem Soc, Dalton Trans (2002) 4307.

Burford N, Spinney H A, Ferguson M J, McDonald R, Chem Comm (2004) 2696.

Cowley A H, Damasco M C, J Am Chem Soc 93 (1971) 6815.

Cowley A H, Kemp R A, Chem Rev 85 (1985) 367.

Ellis B D, Macdonald C L B, Inorg Chem 45 (2006) 6864.

Fleming S, Lupton M K, Jekot K, Inorg Chem 11 (1972) 2534.

Gruber M, Schmutzler R, Schomburg D, Phosphorus, Sulfur and Silicon 80 (1993) 205.

Karacar A, Diplomarbeit, TU Braunschweig (1996).

Karacar A, Freytag M, Thönnessen H, Jones P G, Bartsch R, Schmutzler R, J Organomet Chem 643–4 (2002) 68.

Kaukorat T, Neda I, Schmutzler R, Coord Chem Rev 137 (1994) 53.

Krafczyk R, Ph D Thesis, TU Braunschweig (1998).

Thomas M G, Schultz C W, Parry R W, Inorg Chem 16 (1977) 994.

Vogt R, Jones P G, Schmutzler R, Chem Ber 126 (1993) 1271.

Chapter 7
Transition Metal Complexes

7.1 Phosphanes

Phosphanes possess an electron lone pair with which they can bond to a transition metal. As this bonding interaction results in the transfer of electron density from phosphorus to the metal atom, we would expect a downfield shift in the ^{31}P-NMR spectrum relative to the value for the free ligand that depends both on the σ-donor strength of the phosphane, and the Lewis acidity of the transition metal fragment.

Since most phosphane ligands also show some π-acceptor ability, coordination to a transition metal can result in π-backbonding that would increase the electron density on phosphorus, and thus result in an upfield shift of the signal in the ^{31}P-NMR spectrum. Depending on the magnitude of the downfield shift due to the σ-donicity, and the upfield shift due to the π-acceptor strength of the phosphane, either a net downfield or net upfield shift is observed upon coordination of a phosphane to a transition metal.

How can we estimate the sign and the magnitude of this expected chemical shift? The chemical shift difference observed upon coordination to a transition metal is known as the coordination chemical shift, and defined as:

$$\text{coordination chemical shift: } \Delta\delta = \delta_\text{p} \text{ (complex)} - \delta_\text{p} \text{ (ligand)}$$

Tricoordinate phosphorus ligands are usually strong σ-donors, resulting in downfield coordination shifts of around $\delta_\text{p} = 20\text{–}70$ ppm in the absence of appreciable π-acceptor strength, and depending on the transition metal fragment they coordinate to. Exactly here lies the problem, and we will have to make an educated guess about the π-acceptor strength of our phosphorus ligand and the electronic properties of the transition metal fragment in question. We can do this by drawing on the information already available to us. We know from our discussion of electronic properties of phosphorus ligands (see box story in chapter 5) that the coordination chemical shift is likely to decrease in the following order:

$$\Delta\delta(\text{PR}_3) > \Delta\delta(\text{PAr}_3) > \Delta\delta(\text{P(OR)}_3)$$

We deduce from our knowledge concerning the electronic properties of the d-block metals that the coordination chemical shift is likely to decrease going down the group,

O. Kühl, *Phosphorus-31 NMR Spectroscopy*,
© Springer-Verlag Berlin Heidelberg 2008

$$\Delta\delta(3d) > \Delta\delta(4d) > \Delta\delta(5d)$$

as well as from left to right within the period. However, as the coordination chemical shift depends not so much on the transition metal, but on the electron density that the transition metal possesses, the prediction of the exact position of the chemical shift of a given transition metal phosphane complex based on these considerations alone is perilous. There are simply too many parameters. It would be far easier to look at a specific transition metal fragment, say $[W(CO)_5L]$, and try to develop a method whereby the $\Delta\delta$ of an as yet unsynthesised representative can be calculated from the chemical shift of the free ligand – a value that might be found in the literature. This can easily be done by correlating $\Delta\delta$ with δ_p of the free ligand, and thus arriving at the following equation:

$$\Delta\delta = A \ \delta_p \text{ (free ligand)} + B$$

This equation will have to take both contributing factors (σ-donor ability and π-acceptor strength) into account, and will likely ignore steric and substituent effects, as these are smaller than the two dominating electronic contributions. We would expect that the correlation factor R will be rather small, indicative of the unaccounted additional contributions.

Note: *Accurate correlation equations can only be drawn from similar phosphane ligands, so-called ligand families (see QALE).*

We already know much about the magnitude and the sign of the two parameters A and B, even without computing the equation for an actual transition metal complex. We remember that the part of $\Delta\delta$ attributable to the σ-donicity is large and positive (downfield), whereas the part of $\Delta\delta$ attributable to π-acceptor strength is usually smaller and negative (upfield). In fact, the effect due to the π-acceptor strength increases with the chemical shift of the free ligand. The further downfield the signal for the free ligand, the further upfield the signal for the complex is shifted due to backbonding. This means that A has to have a negative value and carries the ligand influence on the coordination chemical shift. Ligands with a positive phosphorus resonance have a noticeable π-acceptor strength, causing an upfield coordination shift (sign change through A), whereas for ligands with a negative phosphorus resonance, the π-acceptor strength is negligible, resulting in a downfield coordination shift (sign change through A). B is a constant that reflects the effect of coordination via the lone pair on the phosphorus atom, and is therefore positive.

Another important aspect to consider is that π-backbonding from the metal is not the only cause for an upfield shift of the phosphorus signal. Phosphites and phosphorus halides have substituents (O, halogen) capable of hyperconjugation toward the phosphorus atom, resulting in an upfield shift of the phosphorus resonance. It is, therefore, not surprising that these ligands are those with the largest negative coordination chemical shift.

With these considerations in mind, we can now look at actual correlation equations and whether or not they confirm our expectations. A series of these equations are listed in Tables 7.1–7.3, and at a glance, we notice that A is indeed always

Table 7.1 Correlation equations to determine the coordination chemical shift of group 6 carbonyl phosphane complexes

Complex	A	B
$[Cr(CO)_5L]$	−0.153	60.58
$[Mo(CO)_5L]$	−0.142	40.19
$[W(CO)_5L]$	−0.174	19.29
cis-$[Cr(CO)_4L_2]$	−0.196	57.92
cis-$[Mo(CO)_4L_2]$	−0.136	43.90
cis-$[W(CO)_4L_2]$	−0.138	19.35
fac-$[Cr(CO)_3L_3]$	−0.216	64.60
fac-$[Mo(CO)_3L_3]$	−0.105	36.49
fac-$[W(CO)_3L_3]$	−0.094	17.69

Table 7.2 Relative independence of the correlation factors from the ligand count for a given metal

		Cr	Mo	W
A	$[M(CO)_5L]$	−0.153	−0.141	−0.174
	cis-$[M(CO)_4L_2]$	−0.196	−0.136	−0.138
	fac-$[M(CO)_3L_3]$	−0.216	−0.105	−0.094
B	$[M(CO)_5L]$	60.58	40.19	19.29
	cis-$[M(CO)_4L_2]$	57.92	43.90	19.35
	fac-$[M(CO)_3L_3]$	64.60	36.49	17.69

Table 7.3 Dependance of the correlation equation on the metal and the geometry

Complex	A	B
cis-$[Mo(CO)_4L_2]$	−0.136	43.90
cis-$[Ru(CO)_2L_2Cl_2]$	−0.416	26.84
trans-$[Rh(CO)L_2Cl]$	−0.335	35.89
cis-$[PdL_2Cl_2]$	−0.202	38.63
trans-$[PdL_2Cl_2]$	−0.304	26.79

negative and that B decreases down a triad. Closer inspection tells us that we cannot really confirm a horizontal trend in the values for B going from Mo to Pd. If we focus on the relevant equations listed in Table 7.3, the reason becomes immediately apparent. We do not look at a series of complexes that vary only in the central metal atom. Instead, we have different geometries (square planar and octahedral), different ligands (to accommodate the increasing electron count on the metal), and different isomers (cis and trans for phosphorus coordination).

If we consider the series cis-$[Mo(CO)_4L_2]$, cis-$[Ru(CO)_2L_2Cl_2]$, trans-$[Rh(CO)L_2Cl]$, and cis-$[PdL_2Cl_2]$, we find a rather erratic order of B values: 43.90, 26.84, 35.89, and 38.63, that apparently follows no trend at all. The reason, of course, is that the complexes are not closely related in geometry and ligand composition.

If we consider individual pairs that are fairly closely related, then our assumptions are again confirmed.

Example 1 The complexes trans-$[Rh(CO)L_2Cl]$ and trans-$[PdL_2Cl_2]$ have the same square planar geometry and trans-configuration, but differ in the oxidation state

(Rh(I) versus Pd(II)), and in that the Rh(I) complex features a strong π-acceptor coligand (CO), whereas the Pd(II)-complex does not. The difference in oxidation state combined with the difference in the group number results in isoelectronic metal ions (d^8). This would lead us to expect that the difference lies in the coligand, meaning that the available electron density on the metal will be channeled toward the CO-coligand in the Rh(I) case, resulting in a smaller upfield shift due to the π-acceptor strength of the phosphane compared to the Pd(II)-complex. We see this confirmed in the smaller B value for *trans*-[PdL$_2$Cl$_2$].

Example 2 The complexes *trans*-[PdL$_2$Cl$_2$] and *cis*-[PdL$_2$Cl$_2$] differ only in the position of the phosphane ligands relative to each other. We would expect to see a *trans*-influence of the ligands. The more electronegative chloride should compete better for the metal's electron density than the phosphane. Consequently, we should see an increasingly larger upfield shift for increasingly stronger π-acceptor ligands in favour of the *trans*-isomer. In the *trans*-isomer, the phosphane competes with another phosphane, whereas in the *cis*-isomer, the phosphane competes with chloride.

The importance of the *trans*-influence actually depends on the oxidation state of the metal and the coligands. Table 7.5 lists complexes that have the same geometry and possess isoelectronic metal centres. They are all d^6-transition metals. However, in the series *mer*-[M(CO)$_3$L$_3$] (M = Cr, Mo, W), we see that A is small and B is large, indicative of a small effect of backbonding on the chemical shift of the complex. The coordination chemical shift $\Delta\delta$ is driven by the σ-donicity of the ligand. The π-backbonding is mainly diverted into the CO-ligands. However, there is a marked difference between the two different phosphorus nuclei. The two phosphorus atoms mutually *trans* to each other have a significantly smaller B value than the phosphorus atom *trans* to CO. This results in an upfield shift of these two phosphorus atoms relative to the third, as the third loses π-electron density to the better π-acceptor ligand CO *trans* to it.

Table 7.4 Dependance of the coordination chemical shift on the complex geometry

Ligand	δ_P free ligand	$\Delta\delta$ *cis*-[PdL$_2$Cl$_2$]	$\Delta\delta$ *trans*-[PdL$_2$Cl$_2$]	$\Delta\delta_{cis} - \Delta\delta_{trans}$
PMe$_3$	−62	+51	+46	+5
PPh$_3$	−6	+40	+29	+11
PClPh$_2$	+82	+22	+2	+20
P(OPh)$_3$	+128	+13	−12	+25

Table 7.5 Correlation factors for a series of *mer*-complexes

Complex	A	B	Complex	A	B
mer-[Cr(CO)$_3$L$_3$]	−0.111	61.66			
	−0.142	72.71			
mer-[Mo(CO)$_3$L$_3$]	−0.093	35.94	*mer*-[RhL$_3$Cl$_3$]	−0.654	10.36
	−0.091	48.92		−0.741	9.95
mer-[W(CO)$_3$L$_3$]	−0.109	17.33	*mer*-[IrL$_3$Cl$_3$]	−0.799	−30.80
	−0.102	22.33		−0.722	−32.47

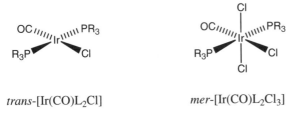

trans-[Ir(CO)L$_2$Cl]　　　　　　　　　*mer*-[Ir(CO)L$_2$Cl$_3$]

Fig. 7.1 Differences in the *trans*- and *mer*-geometries of iridium carbonyl complexes

In the series *mer*-[ML$_3$Cl$_3$] (M=Rh, Ir), we see that A is large (in magnitude) and B is small, indicative of a much smaller influence of σ-donicity on $\Delta\delta$. There is no coligand of appreciable π-acceptor strength present, and thus all of the π-electron density from the metal will be taken by the phosphorus ligands. The difference between the two non-equivalent phosphorus atoms is small, both in A and in B, and we may suspect that the high oxidation state may lower the *trans*-influence on the σ-donicity.

We can verify that statement by looking at a pair of iridium complexes, *trans*-[Ir(CO)L$_2$Cl] with an Ir(I)-center, and *mer*-[Ir(CO)L$_2$Cl$_3$] with an Ir(III)-center. The two complexes differ in that the latter is derived by oxidizing the square planar Ir(I)-complex adding chloride ligands to the axial positions (see Fig. 7.1). Whereas the A value is virtually unaffected by this oxidation (−0.357 and −0.377, respectively), the B value decreases from 29.11 to 0.70 upon going from Ir(I) to Ir(III). In other words, the influence of σ-donicity and thus the *trans*-influence diminishes significantly when the oxidation state of the central metal is increased.

7.2 Chelate Phosphanes (Influence of Ring Size)

We have seen that the ^{31}P chemical shifts in cyclic phosphinidenes are influenced by their position within the ring (and thus by their position with respect to other double bonds), but not necessarily by the ring size as such. Observations from other organophosphorus compounds show that deviations between acyclic and closely-related cyclic systems do exist, but they cannot be explained merely by ring size.

If a bidentate phosphane ligand is coordinated to a transition metal, an additional chemical shift attributable to the size of the resulting metallacycle can be observed for rings containing three, four, five, and six members.

Definition The ring chemical shift Δ_R is defined as the value observed for the metal-lacycle compared to the chemical shift found for a complex where the chelate ligand is replaced by two equivalent monodentate ligands.

An example would be *cis*-[Mo(CO)$_4$dppe] and *cis*-[M(CO)$_4$(PPh$_2$Et)$_2$]. We already know from Sect. 7.1 that coordination of a monodentate phosphane to a transition metal results in an additional chemical shift, compared to the value for the free ligand. We found that this $\Delta\delta$ value was dependant on both the ligand

and the metal. Similarily, the additional chemical shift value attributable to ring size Δ_R is dependant upon ring size, ligand, and metal. Fortunately, the sign varies with the ring size: four- and six-membered rings result in an upfield shift, whereas three- and five-membered rings are shifted downfield. In magnitude, increased ring size results in a smaller Δ_R value. This can be seen from the chemical shift values for 1,ω-bisphosphinoalkyl ligands and their group 6 chelate complexes cis-[M(CO)$_4$L-L] (M=Cr, Mo, W) (see Table 7.6).

From the data in Table 7.6, we can also see that within the Cr, Mo, W triad, Δ_R experiences an upfield shift as one moves down the group, irrespective of ring size. This behaviour follows closely that of the monodentate phosphane ligands, where we have already noticed an upfield shift for the transition metal complexes as one moves down the group. The notable exception is the transition between cis-[Mo(CO)$_4$dppe] and cis-[W(CO)$_4$dppe], where the value of Δ_R increases.

Note: *Four- and six-membered metallacyclophosphanes experience an upfield shift compared to the free ligand.*

Note: *Three- and five-membered metallacyclophosphanes experience a downfield shift compared to the free ligand.*

Note: *The chemical shift value for cis-[W(CO)$_4$dppm] is found to be the same as that for the free ligand dppm. This is because $\Delta_{coord} = -\Delta_R$ for this pair of compounds.*

Note: *The effect is additive. A phosphorus center belonging simultaneously to two metallacycles experiences the sum of the Δ_R values assignable to the individual rings (see Fig. 7.2).*

The dependance of the chemical shift on the ring size is not limited to organophosphorus chelate ligands. It applies to any phosphorus containing metallacycles.

The difference in chemical shift values can be taken as a very good indication for the presence or absence of a metal-metal bond. In a series of palladium complexes with bridging μ-PPh$_2$ groups, the chemical shift value for the four-membered ring (no Pd-Pd bond) is seen ~ $\Delta\delta = -300$ ppm upfield of the chemical shift value for the three-membered ring (with Pd-Pd bond) (see Fig. 7.3).

Table 7.6 Ring-size effect in group 6 metallacycles

M	L-L	δ_P [ppm]	$\Delta\delta$ [ppm]	Δ_R [ppm]	ring size
	dppm	−23.6			
Cr		+25.4	+49.0	−12	4
Mo		0.0	+23.6	−19.6	4
W		−23.7	0	−25.4	4
	dppe	−12.5			
Cr		+79.4	+91.9	+31.0	5
Mo		+54.7	+67.2	+24.3	5
W		+40.1	+52.6	+27.3	5
	dppp	−17.3			
Cr		+41.4	+58.7	−2	6
Mo		+21.0	+38.3	−4.6	6
W		0.0	+17.3	−8.0	6

δ_{Pa} : 89.3 ppm | δ_{Pa} : 56.0 ppm | δ_{Pa} : 18.8 ppm

δ_{Pb} : 69.7 ppm | δ_{Pb} : 68.9 ppm

δ_{Pc} : 30.2 ppm | δ_{Pc} : 29.2 ppm

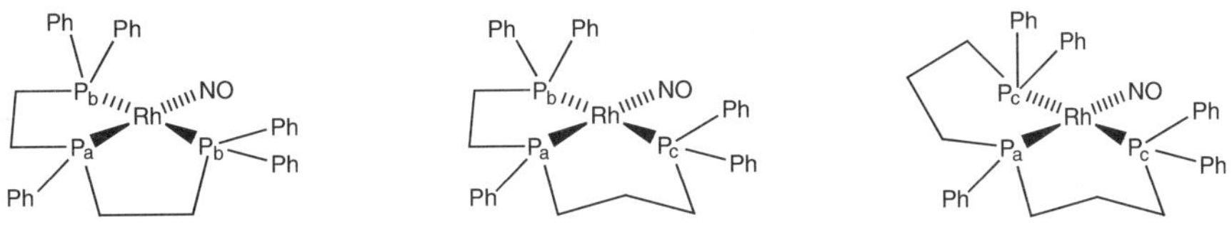

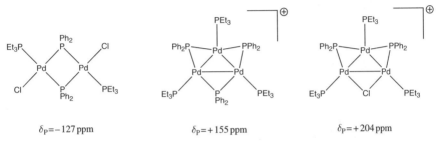

Fig. 7.2 Additive effect of ring incorporation in phosphorus containing metallacycles

$\delta_P = -127$ ppm | $\delta_P = +155$ ppm | $\delta_P = +204$ ppm

Fig. 7.3 Dependance of the phosphorus resonance on the existence of a metal-metal bond (three-membered vs. four-membered metallacycle)

7.3 Phosphides (Terminal and Bridging)

Here, we are faced with phosphorus resonances that are frequently very far downfield indeed. Most of these phosphorus atoms are subject to very large shift anisotropies, making accurate predictions and simple explanations difficult. However we are not concerned with accurate predictions. We are looking for an approximate chemical shift value and the best advice for the practitioner in this context is to ask the NMR operator to open the window a few hundert ppm further to the left.

We will first discuss the influence of coordination on the chemical shift values of bridging phosphanido ligands in a series of molybdenum and tungsten complexes of the general formulae $[\{CpM(CO)_2\}_2(\mu\text{-}PRR')_2]$ (see Table 7.7), $[\{CpM(CO)\}_2(\mu\text{-}PRR')_2]$ (see Table 7.9 and Fig. 7.4) and $[\{CpM\}_2(\mu\text{-}PRR')_2(\mu\text{-}CO)]$ (see Table 7.8) with varying degrees of metal-metal bonding. As expected, the phosphorus resonances in the tungsten complexes appear upfield from those of the respective molybdenum complexes.

As the structures have all been assigned a metal-metal bond by the original authors and thus possess a three-membered M_2P metallacycle, we would expect the chemical shifts to be in the downfield region at around $\delta_P = 150\text{–}300$ ppm for

Table 7.7 Chemical shift values for the [{CpM(CO)$_2$}$_2$(μ-PRR')$_2$] series

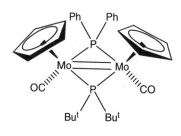

R	R'	Mo, δ_P	W, δ_P
Cy	Cy	218.8	139.6 (182)
Et	Et	179.8	99.6 (197)
Cy	H	152.6	
Ph	Ph		109.2 (209)

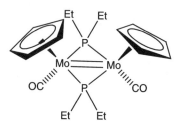

δ_P: 213.0 ppm (PBu$^t_2$)
148.0 ppm (PPh$_2$)

δ_P: 194.7 ppm

Fig. 7.4 ^{31}P-NMR chemical shift values for *cis*-[{CpM(CO)}$_2$(μ-PRR')$_2$]

Table 7.8 Chemical shift values for the [{CpM}$_2$(μ-PRR')$_2$(μ-CO)] series

R	R'	Mo, δ_P	W, δ_P
Cy	Cy	263.7	
Et	Et	212.3	147.1 (376) toluene-d8
Cy	H	167.3	
Ph	Ph		144.7 (389) CD$_2$Cl$_2$

molybdenum, and δ_P = 100–200 ppm for tungsten. This is indeed the case for the [{CpM(CO)$_2$}$_2$(μ-PRR')$_2$] and [{CpM}$_2$(μ-PRR')$_2$(μ-CO)] series.

Note: *The [{CpM}$_2$(μ-PRR')$_2$(μ-CO)] series has phosphorus resonances that are deshielded by $\Delta\delta$ = 15–45 ppm, compared to the respective values in the [{CpM(CO)$_2$}$_2$(μ-PRR')$_2$] series. This is attributed to the higher metal-metal bond order in the former complexes.*

The case of [{CpM(CO)}$_2$(μ-PRR')$_2$] is not that straightforward. There are two isomers, one in which both Cp-ligands are on one side of the M-M vector and both the carbonyl groups on the other, termed the *cis*-isomer (see Fig. 7.4), and the other isomer is *trans*-[{CpM(CO)}$_2$(μ-PRR')$_2$] with the Cp-ligands on opposite sides of the M-M vector (see Table 7.9).

The respective phosphorus resonances are dramatically different with the chemical shifts in *trans*-[{CpM(CO)}$_2$(μ-PRR')$_2$] shielded by about $\Delta\delta = 100$ ppm, compared to *cis*-[{CpM(CO)}$_2$(μ-PRR')$_2$]. There is no reason why the relative positions of the coligands to the phosphanides should result in such a large chemical shift difference, especially as there is no *trans*-influence in tetrahedral complexes. Furthermore, the region of $\delta_P = 0–100$ ppm is the region where we would have to expect the resonance of a bridging phosphanide ligand in the absence of an M-M bond. Indeed, the observed downfield shift from *trans*-[{CpM(CO)}$_2$(μ-PRR')$_2$] to *cis*-[{CpM(CO)}$_2$(μ-PRR')$_2$] is suggestive of the absence of M-M bonds in the *trans*-isomer.

We conclude that the assignment of a M = M double bond in trans-[{CpM(CO)}$_2$ *(μ-PRR')2] is not in agreement with the observed chemical shift values in the* ^{31}*P-NMR spectrum, and largely due to a desire to fulfill the 18 valence electron rule.*

The coordination of primary phosphanides to a Cp$^R_2$ZrCl-fragment results in a significant downfield shift of the phosphorus resonance of some $\Delta\delta = 150–220$ ppm relative to the value of the primary phosphane, and, dependant on the nature of the Cp-ligand and the substituents on phosphorus, a surprisingly large deshielding when compared to the values of bridging phosphanides in the absence of metal-metal bonds (see Fig. 7.5).

If the chloride ligand is substituted by a second phosphanide, an upfield shift of $\Delta\delta = -38.8$ ppm is observed in the reaction of [Cp$_2$ZrCl(PHSmes)] to [Cp$_2$Zr(PHSmes)$_2$].

Another interesting aspect is the fact that the phosphorus NMR spectra of the complexes [Cp$_2$Zr(PHR)$_2$] (R = Bu^t, Smes) show different coupling patterns (see Fig. 7.6). While both exhibit a singlet in the ^{31}P-{^{1}H}-NMR spectrum, the ^{31}P-NMR spectrum of [Cp$_2$Zr(PHBu^t)$_2$] features a doublet ($^1J_{PH} = 209$ Hz), but that of [Cp$_2$Zr(PHSmes)$_2$] displays a multiplet reminiscent of a doublet of triplet, but was interpreted by the authors as a doublet of doublets of doublets ($^1J_{PH} = 247$ Hz, $^2J_{PP} = 50$ Hz, $^3J_{PH} = 5$ Hz), revealing the inequivalence of the two phosphorus atoms. The signal is likely the result of a dynamic process between a pyramidal and a

Table 7.9 ^{31}P-NMR chemical shift values for *trans*-[{CpM(CO)}$_2$(μ-PRR')$_2$]

R	R'	Mo, δ_P	W, δ_P
Cy	Cy	95.4	
Et	Et	78.7	16.7 (291)
Cy	H	45.4	
Ph	Ph		34.7 (301)

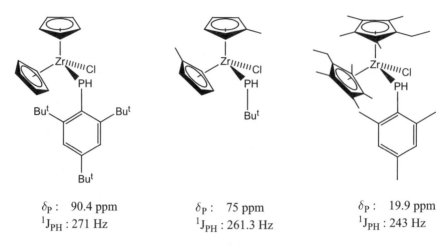

δ_P :   90.4 ppm δ_P :   75 ppm δ_P :   19.9 ppm

$^1J_{PH}$: 271 Hz $^1J_{PH}$: 261.3 Hz $^1J_{PH}$: 243 Hz

Fig. 7.5 ^{31}P-NMR chemical shift values for [Cp$^R_2$ZrCl(PHR')]

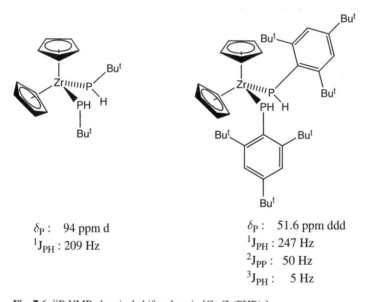

δ_P :   94 ppm d

$^1J_{PH}$: 209 Hz

δ_P :   51.6 ppm ddd

$^1J_{PH}$: 247 Hz

$^2J_{PP}$: 50 Hz

$^3J_{PH}$:   5 Hz

Fig. 7.6 ^{31}P-NMR chemical shift values in [Cp$_2$Zr(PHR)$_2$]

planar PR$_2$-fragment, and thus not first order. With the smaller phosphanide, the process is faster and results in a singlet on the NMR timescale.

Evidence comes from the x-ray crystal structure determination of [Cp$_2$Zr (PHSmes)$_2$] and [Cp$_2$Hf{P(SiMe$_3$)$_2$}$_2$], where two different geometries were found for the two phosphorus atoms. One phosphorus atom is planar (three electron donor), the other pyramidal (one electron donor). Variable temperature phosphorus NMR was performed on a closely-related complex, [Cp$_2$Hf(PCy$_2$)$_2$], showing two resonances at $\delta_p = 270.2$ ppm and $\delta_p = -15.3$ ppm at $-126°C$. The deshielded signal belongs to the π-bonded, planar, three electron donor $=PCy_2$ whereas the shielded signal is assigned to the σ-bonded, pyramidal, one electron donor $-PCy_2$. At room temperature, exchange between the two bonding modes is fast on the NMR time scale, and an average signal is observed with an intermediate chemical shift value.

Having found the reason for the surprisingly large downfield shift in the terminal phosphanide complexes of the group four metals, it is not surprising that the complexes with bridging phosphanides have very similar chemical shifts. The complexes $[Cp'_2Zr(\mu\text{-}PHR)]_2$ (R = Ad, Bu^t) are four-membered metallacycles without metal-metal bonds, and as such should show an upfield ring shift value (see Fig. 7.7). This explains the relatively shielded resonances at about $\delta_p = 125$ ppm. Another difference is the geometry on phosphorus. The terminal three electron donor phosphanide $=PR_2$ is trigonal planar and deshielded at $\delta_p = 270$ ppm, whereas the bridging three electron donor phosphanide $-PR_2-$ is pyramidal and relatively shielded at $\delta_p = 125$ ppm.

Note: *The geometry around phosphorus is an important factor influencing the chemical shift value. Trigonal planar phosphorus atoms resonate downfield from pyramidal ones.*

Substitution of the second Cp_2Zr-unit by a PR-fragment does not have a large effect on the chemical shift of the metal-bound phosphorus atoms (see Table 7.10 and Fig. 7.8). The P_3-unit displays an A_2X pattern with the A_2 resonances at $\delta_p = 92-138$ ppm and the X resonance at $\delta_p = -156$ to -190 ppm, indicative of π-bonding interactions of the $M\text{-}P_{A2}$ bonds and the absence of bonding of P_x to the metal. Comparison of the respective chemical shift values down the group (M = Ti, Zr, Hf) again shows the general trend that the phosphorus resonance becomes more shielded moving down the triad. Naturally, this trend is far less pronounced in P_x due to the absence of direct $M\text{-}P_x$ bonding.

Comparison of the chemical shift values in $[Cp°_2Zr(\eta^2\text{-}P_2Mes_2)]$ $\delta_p = 138.7$ ppm and $[Cp'_2Zr(\eta^2\text{-}P_3Bu^t_3)]$ $\delta_p = 139.0$ ppm (P_{A2}) gives the impression that the electron lone pairs of the phosphorus atoms in the ZrP_2 three-membered ring are unsuited

δ_P: 124.6 ppm d δ_P: 126.7 ppm d
$^1J_{PH}$: 251 Hz $^1J_{PH}$: 258.7 Hz

Fig. 7.7 ^{31}P-NMR chemical shift values for $[Cp'_2Zr(PHR)_2]$

Table 7.10 ^{31}P-NMR chemical shift values for $[Cp_2Zr(\kappa^2-P_3R_3)]$

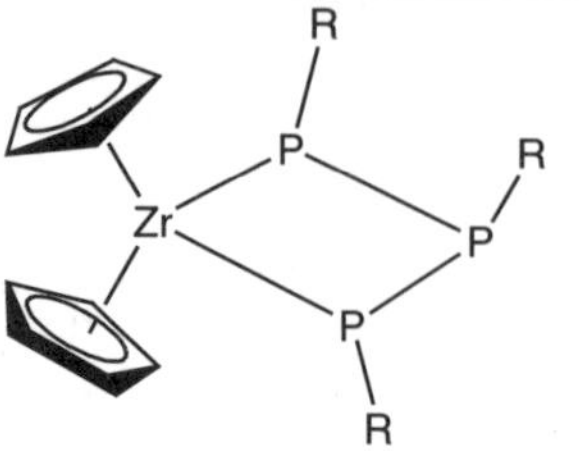

	Metal	A$_2$ [ppm]	X [ppm]
	Ti	182–154	−143/−187
	Zr	92–138	−156/−190
	Hf	67–106	−160/−195

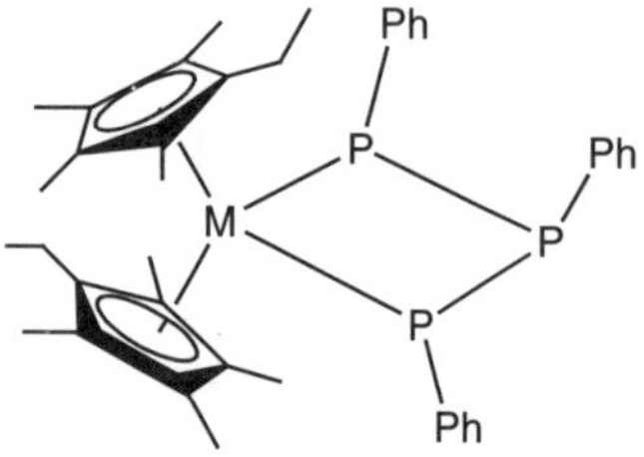

	M	δ_P A2 [ppm]	δ_P X [ppm]
	Zr	60.87	−134.89
	Hf	36.37	−128.75

Fig. 7.8 ^{31}P-NMR chemical shift values for $[Cp^\circ_2Zr(\kappa^2-P_3Ph_3)]$

to engage in π-bonding interactions with appropriate metal orbitals on the zirconium atom (see Fig. 7.9). A three-membered metallacycle, as in ZrP$_2$ results in a large downfield ring shift, whereas the four-membered metallacycle in ZrP$_3$ would be responsible for a considerable upfield ring shift. That the two compounds have almost identical phosphorus resonances, despite the opposing ring shifts, must have its reason in different M-P bonds.

In these zirconium complexes, the zirconium center has formally 16 VE, but can achieve 18 VE if it is the recipient of a π-donor interaction from either of the two phosphorus substituents possessing a lone pair. Whereas this π-bonding interaction is evidently possible in $[Cp'_2Zr(\eta^2-P_3R_3)]$ (R = Ad, Bu^t), it is prevented by the ring

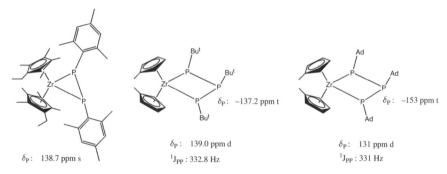

Fig. 7.9 Comparison of ^{31}P-NMR chemical shifts in $[Cp^\circ_2Zr(P_2Mes_2)]$ and $[Cp'_2Zr(\kappa^2-P_3R_3)]$

strain within the ZrP$_2$ three-membered ring in [Cp°$_2$Zr(η^2-P$_2$Mes$_2$)]. It is interesting to note that the effect of the π-bonding in [Cp'$_2$Zr(η^2-P$_3$R$_3$)] incidentally equals the $\sum \Delta_R$ for the three- and four-membered metallacycles.

In a series of iron complexes with bridging phosphanido ligands (see Fig. 7.10), we again observe the phosphorus resonances at $\delta_P = 100$–200 ppm, indicative of a Fe-Fe bond, and thus involvement of the phosphorus atom in Fe$_2$P three- membered metallacycles with the associated downfield ring shift. In the pair [{Fe(CO)$_3$}$_2$ {μ-o-(PPh)$_2$-C$_6$H$_4$}] $\delta_P = 136.2$ ppm and [(CpFe)$_2$(μ-CO)(μ-PPh$_2$)$_2$] $\delta_P\delta_P = 116.9$ ppm, we can see that the substitution of the strong π-acceptor ligand CO with a Cp-ligand results in a noticeable upfield shift, as expected.

In going from secondary to primary phosphanides, the expected upfield shift in the phosphorus resonance is observed as primary phosphanes are usually found upfield from the respective secondary phosphanes. Substituting a phenyl group with a hydrogen atom does not only create an upfield shift in the phosphorus resonance, but also introduces asymmetry into the complex (see Fig. 7.11). There are now three stereoisomers observable. They are differentiated by the relative positions of the phenyl and hydrogen substituents on the phosphorus atoms. Each of the three isomers has its characteristic multiplet pattern, with accompanying J$_{PH}$ and J$_{PP}$ coupling constants.

Note: *The isomer with both hydrogen atoms in axial positions displays the same multiplet and coupling constants as [Cp$_2$Zr(PHSmes)$_2$], making it possible to assign a structure to the compound. [Cp$_2$Zr(PHBu^t)$_2$] has the same pattern as the isomer with both phenyl groups in axial positions.*

Substitution of two carbonyl groups (four electrons) with a Cp-ligand (five electrons) results in the disappearance of the Fe—Fe bond (see Fig. 7.12). In consequence, we expect a dramatic upfield shift as the Fe$_2$P three-membered metallacycle is replaced by a Fe$_2$P$_2$ four-membered metallacycle. The phosphorus resonances in [CpFeCO(μ-PPh$_2$)]$_2$ are observed at $\delta_P = -8.1$ ppm (*cis*) and $\delta_P = -11.6$ ppm (*trans*), respectively, some $\Delta\delta = -125$ ppm upfield from the closely related complex [(CpFe)$_2$(μ-CO)(μ-PPh$_2$)$_2$].

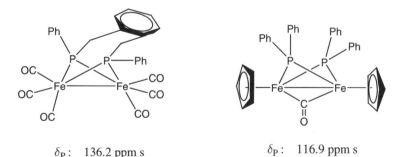

δ_P: 136.2 ppm s δ_P: 116.9 ppm s

Fig. 7.10 Substitution of CO ligands by a Cp-ligand

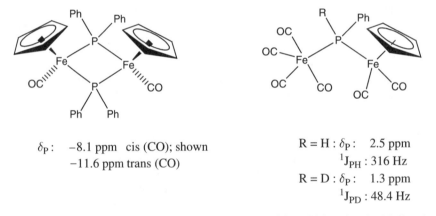

Fig. 7.11 The three isomers of [{Fe(CO)$_3$}$_2$(μ-PHPh)$_2$]

δ_P: −8.1 ppm cis (CO); shown
−11.6 ppm trans (CO)

R = H : δ_P: 2.5 ppm
$^1J_{PH}$: 316 Hz
R = D : δ_P: 1.3 ppm
$^1J_{PD}$: 48.4 Hz

Fig. 7.12 Comparison of a cyclic and an acyclic complex with a bridging phosphanide ligand

Removal of the second μ-PPh$_2$ group (three CO ligands introduced for five electrons from Cp and one from PPh$_2$ on one iron atom, and substitution of CO for PPh$_2$ on the other) yields the acyclic bridging phosphanide complex [Fe(CO)$_4$(μ-PEPh)FeCp(CO)$_2$] (E=H, D) (see Fig. 7.12), with a phosphorus resonance of δ_P=2.5 ppm (E=H) and δ_P=1.3 ppm (E=D). Our expectation was to see a downfield shift for the disappearance of the four-membered metallacycle and an upfield shift for the substitution of a phenyl group by a hydrogen atom. As it turns out, the two effects cancel each other, resulting in only a very moderate downfield shift of about $\Delta\delta$=10 ppm.

Note: *There is a pronounced isotope effect on the coupling constant. The $^1J_{PH}$ coupling constant (316 Hz) is one order of magnitude larger than the $^1J_{PD}$ coupling constant (48.4 Hz). We observe a reduced coupling constant for deuterium.*

Substitution of a bridging phosphanide with a bridging hydride in [CpFe-CO (μ-PPh$_2$)]$_2$ (see Fig. 7.13) reintroduces the Fe—Fe bond, accompanied by a dramatic downfield shift of about $\Delta\delta = 200$ ppm due to the replacement of a four-membered metallacycle with a three-membered metallacycle.

Note: *There is no appreciable difference in the phosphorus resonances of the* cis- *and* trans-*isomers.*

An interesting example of a bridging phosphanide is presented by [Fe(CO)$_3$ {μ-(PPhPPh$_2$)}]$_2$,where the bridging phosphanide is, in fact, a diphosphanide (see Fig. 7.14). The bridging phosphorus atom resonates $\Delta\delta = -20$ ppm upfield from the closely related complex [{Fe(CO)$_3$}$_2$\{μ-o-(PPh)$_2$-C$_6$H$_4$\}], a consequence from the PPh$_2$-substituent that is usually responsible for a small upfield shift compared to a simple R-group.

The properties of benzazaphospholes as free ligands were discussed in Sect. 5.2. We need to revisit them to discuss their ^{31}P-NMR chemical shifts when bonded to transition metals. They provide an interesting case study for ligands experiencing a negative coordination chemical shift without formation of a metallacycle.

In Table 7.11, the phosphorus resonances of some group 6 carbonyl complexes of benzazaphospholes are shown alongside their coordination chemical shifts where known. Even for the chromium complex, the coordination chemical shift is only

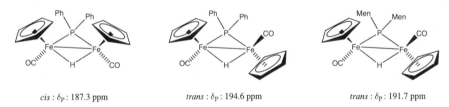

cis : δ_P: 187.3 ppm   trans : δ_P: 194.6 ppm   trans : δ_P: 191.7 ppm

Fig. 7.13 ^{31}P-NMR chemical shift values in *cis*- and *trans*-[(CpFeCO)$_2$(μ-H)(μ-PPh$_2$)]

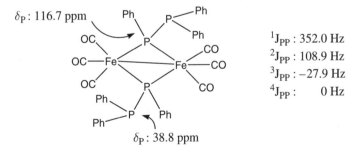

δ_P: 116.7 ppm

$^1J_{PP}$: 352.0 Hz
$^2J_{PP}$: 108.9 Hz
$^3J_{PP}$: −27.9 Hz
$^4J_{PP}$:   0 Hz

δ_P: 38.8 ppm

Fig. 7.14 ^{31}P-NMR chemical shifts in a bridging diphosphanide complex

Table 7.11 ^{31}P chemical shift values for selected metal coordinated benzazaphospholes

R	R'	M	δ_P [ppm]	$^1J_{PC}$ [Hz]	$\Delta\delta$ [ppm]
Me	H	Cr	79.8	22.2	10.0
Bu^t	H	Cr	70.0	7.4	
Me	H	Mo	64.7	19.8	−5.1
Me	H	W	36.7	30.8	−33.1
Bu^t	H	W	24.5	35.1	
Ph	H	W	39.8	29.2	
Me	Me	W	34.8	30.7	−37.0

small, even if downfield. For the higher homologues, the coordination chemical shift is upfield, with the tungsten complexes showing a significant upfield shift of $\Delta\delta=-33$ to -37 ppm. We have already noticed that the nitrogen lone pair in the free ligand engages in a π-donor interaction toward the neighbouring carbon atom in the ring, causing an upfield shift in the phosphorus resonance. The benzazaphosphole can be regarded as an aromatic heterocycle, and is indeed isoelectronic to benzimidazole. This effect is even more pronounced in the deprotonated benzazaphospholates, where the negative charge is located on phosphorus rather than on nitrogen. The same effect is operative here. When the phosphorus atom coordinates to the transition metal fragment using its lone pair, the phosphorus atom becomes electron deficient and suffers a downfield shift. As a means of compensation, the bond order in the P=C bond decreases and that in the N-C bond increases, resulting in an upfield shift of the phosphorus resonance. The group 6 triad shows the accustomed effect in chemical shifts, meaning an upfield movement of the phosphorus resonance as one goes down the list. As described previously, that has its reason in the additional shielding of the positive core charge by the inner d- and f-shells against the valence shell seen for the higher homologues molybdenum, and especially tungsten.

Whereas the complexes in Table 7.11 still keep a formal but weakened P=C double bond, this double bond has been dissolved in favor of a σ-donor bond toward nickel in the dimeric [CpNi(μ-benzazaphospholate)]$_2$ complexes depicted in Table 7.12.

The most striking feature is the large upfield shift in their phosphorus resonances, leading to values of $\delta=-96.5$ to -156.0 ppm, indicative of a complete absence of double bonds involving phosphorus. The effect of substitution on the benzazaphosphole

Table 7.12 ^{31}P chemical shift values for a series of nickel-coordinated benzazaphospholes

R	R'	δ_P [ppm]
Me	H	−131.7
Bu^t	H	−155.4; −154.2
Ph	H	−141.6; −144.0
Me	Me	−96.5
Bu^t	Me	−156.0
Ph	Me	−144.7

ring system is interesting to note. As expected, the effect is most pronounced in the substituent on the five-membered benzazaphosphole ring, and only marginal with substitution on the annelated benzene ring. The phosphorus resonance follows the established I/M-effects in aromatic and heteroaromatic ring systems. The reported downfield shift for the couple Me/H to Me/Me raises doubts as to whether the resonance of the latter (Me/Me) was reported correctly by the original authors.

Note: *The substituent R operates on a heteroaromatic ring system, and not on the phosphorus atom directly. The effect of substitution thus follows the rules for aromatic ring systems, and not those for phosphanes (difference in electronegativity of carbon and phosphorus).*

In Fig. 7.15 we see the same structure, but with a simple phosphole rather than a benzazaphosphole ligand. The *pentahapto* bonded phosphole ligand replacing the more standard Cp-ligand is isoelectronic and isoster to it, and thus does not change the general picture. Likewise, the μ-P bridging propholate ligands are isoelectronic to the μ-P bridging benzazaphospholate ligands in Table 7.12. After acknowledging these facts, it should not be surprising that the respective phosphorus resonances are similar. The pholate resonance is seen somewhat downfield from that of the benzazaphospholate.

If the phosphole is replaced by a phosphanide, the phosphorus resonance is shifted upfield by some $\Delta\delta = -100$ ppm. We have previously (Fig. 7.3) seen an example for a phosphanide-bridged palladium(II) system featuring 16 VE and a chemical shift of $\delta_p = -127$ ppm, very similar to the $\delta_p = -153$ to -168 ppm seen in the complexes shown in Table 7.13. In both complexes, the phosphorus atoms are embedded in a four-membered metallacycle responsible for an upfield ring shift, but the nickel(II) complex has 18 VE on the metal, making it more electron rich than the 16 VE palladium(II) complex.

The phosphanido-bridged CpNi complexes exist in two isomers, denoted *syn* and *anti*. In the *syn*-isomer, both phosphorus atoms are found on the same side of the

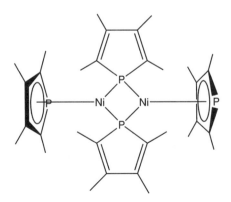

Fig 7.15 ^{31}P chemical shift values in [Ni$_2$\{P(CMe)$_4$\}$_4$]

δ_P: 11.3 ppm, t, J_{PP} = 3.5 Hz (η-C$_4$Me$_4$P)
−118.8 ppm, t, J_{PP} = 3.5 Hz (μ-C$_4$Me$_4$P)

Table 7.13 ^{31}P chemical shift values in a series of nickel-coordinated bridging secondary phosphanides

anti:

syn:

	R$_1$	R$_2$	isomer	δ_P [ppm]	$^2J_{PP}$ [Hz]	$^1J_{PH(D)}$ [Hz]
	H	H		−305	440	303
	H(D)	H	*anti*	−237	432.6	318.3 (48.9)
	H(D)	H	*syn*	−220.2	364.9	326.8
	H(D)	Me	*anti*	−168.4	425.6	315 (48.4)
				−223.6 (H)		
	H(D)	Me	*syn*	−153.7	373.8	318 (48.9)
				−212.2 (H)		
	Me	Me	*anti*	−168		
	Me	Me	*syn*	−154.6		

Ni-Ni vector, whereas in the *anti*-isomer, they are found on opposite sides. In the ^{31}P-NMR spectrum, the resonances for the *syn*-isomer are found $\Delta\delta = 10$–20 ppm downfield from the resonance for the *anti*-isomer.

Note: *The $^1J_{PD}$ coupling constant for the deuterium compound is about one order of magnitude smaller than the $^1J_{PH}$ coupling constant for the ^1H compound.*

Note: *Replacement of the methyl group on phosphorus with a hydrogen atom causes an upfield shift of $\Delta\delta = -60$ ppm, whereas replacement of the hydrogen atom by deuterium does not cause a significant shift of the phosphorus resonance.*

If the primary phosphanide complexes depicted in Table 7.13 are deprotonated, they experience a downfield shift of their phosphorus resonances, as seen in Fig. 7.16. This is rather unexpected, and merits a closer look. The deprotonated phosphorus atom would acquire a lone pair, and thus an increase in its electron density. This is usually accompanied with an upfield shift. What is seen, however, is a downfield shift, indicating a π-donor interaction of the phosphorus atom towards the electron rich nickel atom. In the neutral molecule, the nickel atom is already a 18 VE fragment and a π-donor interaction from the deprotonated phosphanide center would turn it into a 20 VE fragment, a highly unlikely supposition.

There is an alternative explanation (see Fig. 7.17). We initially have anionic, reductive phosphorus atoms in the immediate vicinity of Ni(III) centers. Electron transfer from phosphorus to nickel results in the reduction of Ni(II) to Ni(I), and unpaired electrons on nickel and phosphorus. Formation of a Ni-Ni bond is feasible

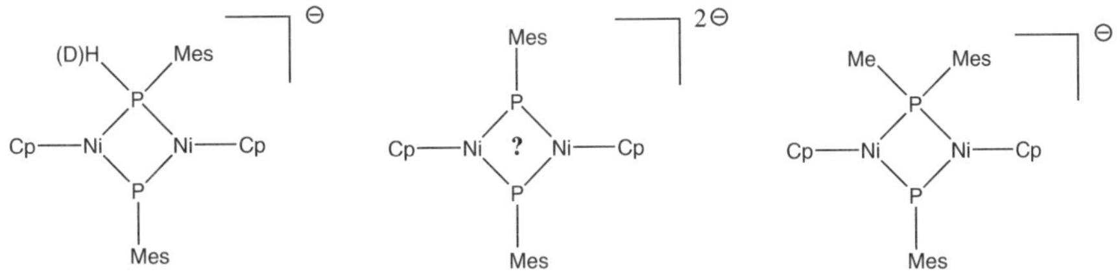

δ_P:	−150.8 ppm	13.8 ppm	−99.3 ppm
	−94.2 ppm		−56.9 ppm
$^2J_{PP}$:	198 Hz		198 Hz
$^1J_{PH}$:	296 Hz		
$^1J_{PD}$:	45.8 Hz		

Fig 7.16 ^{31}P chemical shift values for some related nickel bridging primary and secondary phosphanides

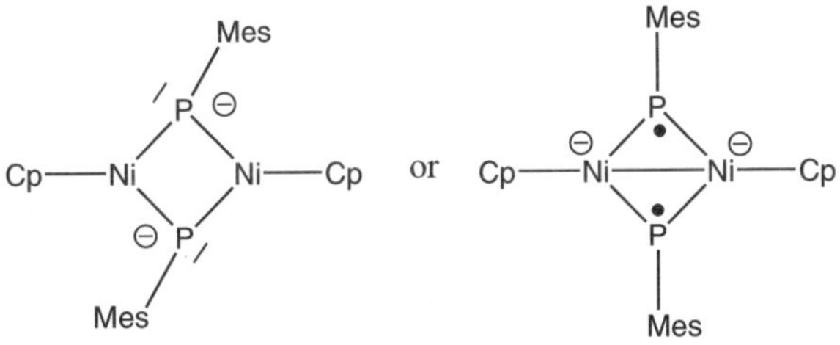

Fig 7.17 The two alternative structures of $[\{CpNi(\mu\text{-}PMes)\}_2]^{2-}$

and observed in similar circumstances, creating two Ni$_2$P three-membered rings that show the observed downfield shift of the phosphorus resonances.

Note: *We have reached the limits of structural determination using ^{31}P-NMR spectroscopy. We are faced with two possible structures that are equally valid, but unsatisfactory explanations for their existences.*

7.4 Phosphinidines (Terminal and Bridging)

Primary phosphanes, like primary amines, possess two hydrogen atoms and thus can be doubly deprotonated. A doubly deprotonated primary phosphane is known as a phosphinidine. It can realize a variety of different bonding modes with metals (see Fig. 7.18). It can form two covalent bonds and one donor bond. It can act as a two electron or a four electron donor ligand, either as a terminal or a bridging ligand.

The complex depicted in Fig. 7.19 features an extremely downfield shifted phosphorus resonance at $\delta_p = 1362$ ppm, and has a bonding pattern that is

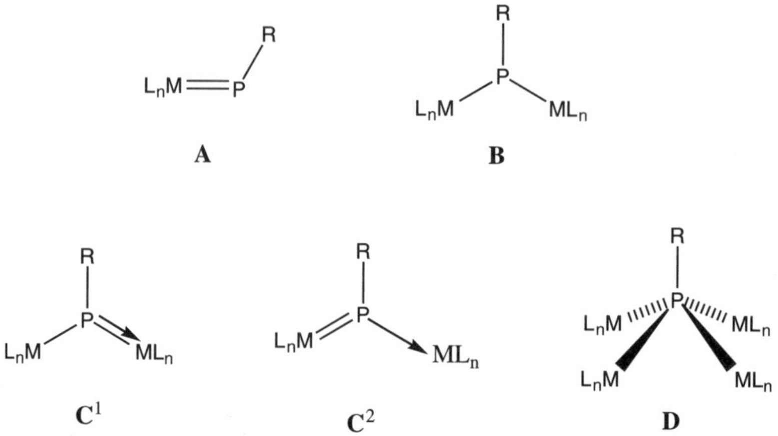

Fig. 7.18 Bonding modes in phosphinidene metal complexes

Fig. 7.19 The ^{31}P chemical shift in [{Cr(CO)$_5$}$_2$(μ-PBu^t)]

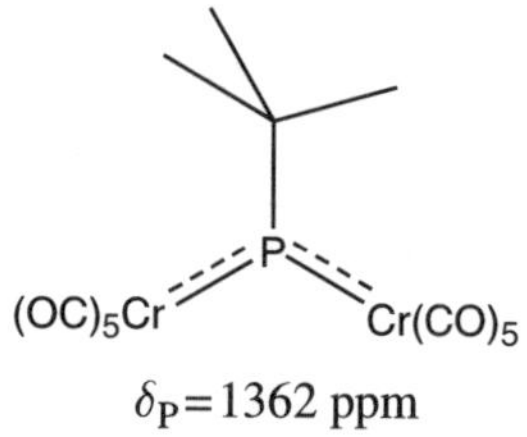

$\delta_P = 1362$ ppm

essentially of type C^1 (see Fig. 7.18), with a trigonal planar geometry around phosphorus. We are familiar with metal complexes of the type [Cr(CO)$_5$L], where L is a two electron σ-donor ligand. However, if we examine the bonding in [{Cr(CO)$_5$}$_2$(μ-PBu^t)] more closely, we find that the compound is not our standard [Cr(CO)$_5$L] complex. We should rather view it as a combination of a Bu^{t}P^{2-} entity featuring three lone pairs and two Cr(CO)$_5^+$ moieties containing a chromium(I) center, and being a 15 VE fragment rather than the 16 VE fragment encountered in [Cr(CO)$_5$L].

The 15 VE Cr(CO)$_5^+$ fragment requires a further three electrons to become an 18 VE complex. Each Cr(CO)$_5^+$ moiety will acquire two electrons from a σ-donor bond contributed by a phosphorus lone pair. The remaining third lone pair at phosphorus then has to be shared in two π-interactions with both chromium(I) centers, creating the observed highly unusual downfield resonance at $\delta_p = 1362$ ppm, one of the largest phosphorus chemical shift values ever measured.

An alternative view of the bonding situation in [{Cr(CO)$_5$}$_2$(μ-PBu^t)] would formulate a chromium-chromium bond, placing the phosphorus atom into a three-membered metallacycle that would cause a substantial downfield shift. However, the chromium atoms would then no longer require a π-donor interaction involving the phosphorus lone pair, and the lack of π-interactions on the phosphorus center makes the experimentally observed downfield shift of $\delta_P = 1362$ ppm less plausible. This is even more intriguing as similar complexes featuring metal-metal bonds resonate some $\Delta\delta = -700$ ppm upfield.

In Fig. 7.20, we see the two phosphorus atoms as four electron donors occupying *trans* apexes of an octahedron, with the other positions filled by cobalt(I) centers. An electron count reveals that the cobalt(I) moieties are 14 VE fragments with four singly occupied border orbitals of similar symmetry and orientation as the four singly occupied border orbitals of phosphorus. In short, the cobalt(I) fragments are isolobal to the phosphorus unit.

The phosphorus resonances are found at $\delta_P = 156$ ppm and 257 ppm, respectively, and thus considerably upfield from where we might expect them, seeing that each phosphorus atom is embedded in four three-membered metallacycles. Obviously, this M_4P_2 octahedron constitutes a "special structure" similar to the M_nP_{4-n} tetrahedra that we will encounter in Sect. 7.7 when we discuss "naked" phosphorus ligands. In these "special structures", the symmetry, orientation, and number of the border orbitals of the metal and phosphorus are perfectly matched for the polyhedron whose vertices they occupy. Thus, in octaheral geometries, we see 14 VE metal fragments and four electron phosphorus donors (forming four single bonds, no donor bonds), whereas in the tetrahedral structures we will see 15 VE metal fragments and three electron phosphorus donors (forming three single bonds, no donor bonds, but retaining a stereoactive lone pair).

In Fig. 7.21, we encounter the same situation that was puzzling us while discussing the extreme downfield shift in the phosphorus resonance of $[\{Cr(CO)_5\}_2 (\mu\text{-}PBu^t)]$, only that the chemical shift value can be found some $\Delta\delta = -700$ ppm upfield in a region that we are more comfortable with, given the electronic structure of the compounds.

From an electronic point of view, three carbonyl ligands provide six electrons, the same number as one Cp-ligand and a metal-metal bond combined (if you want

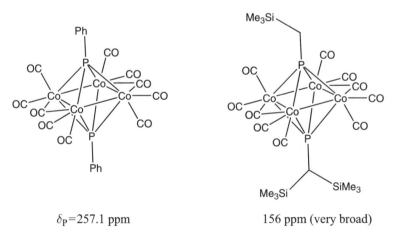

$\delta_P = 257.1$ ppm   156 ppm (very broad)

Fig. 7.20 ^{31}P chemical shift values of two cobalt complexes with the phosphinidenes in *trans* apexes of a Co_4P_2 octahedron

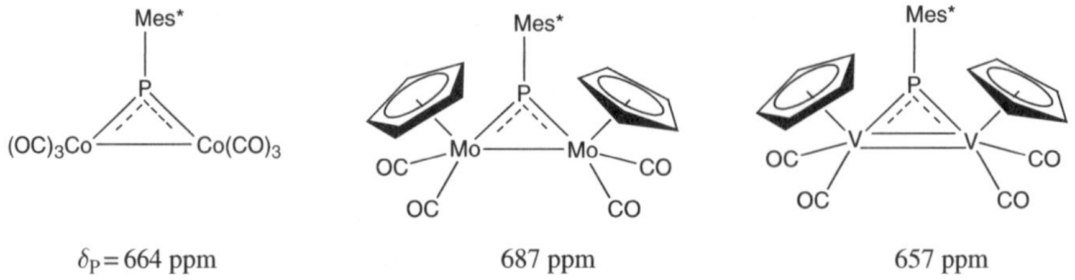

$\delta_P = 664$ ppm   687 ppm   657 ppm

Fig. 7.21 ^{31}P chemical shift values of a series of complexes with a bridging phosphinidene ligand

to count it ionic, then Cp$^-$ is a six electron donor, but the sixth electron comes from molybdenum and thus the net count would still be five electrons donated to the metal). That means that the molybdenum (II) complex featured in Fig. 7.21 with a metal-metal bond should present the same electronic situation at phosphorus as the chromium(I) complex shown in Fig. 7.19 without the metal-metal bond. That makes the chromium complex inconsistent with respect to its phosphorus resonance.

The chemical shift value for [{CpMo(CO)$_2$}$_2$(μ-PMes*)] at $\delta_p = 687$ ppm is in accord with other similar compounds like [{Co(CO)$_3$}$_2$(μ-PMes*)] at $\delta_p = 664$ ppm, and [{CpV(CO)$_2$}$_2$(μ-PMes*)] at $\delta_p = 657$ ppm. The two major contributions to the downfield shift are the three-membered metallacycle, and the π-donor interaction of the phosphorus lone pair with both metal centres. This is not surprising, as the PMes* moiety is identical in all three compounds, and the metal fragments have 15 VE (Co and Mo) and 14 VE (V), respectively. They are isoelectronic with respect to the phosphorus moiety, and thus the vanadium centres are allowed to make up their "formal electron deficiency" by upgrading the metal-metal single bond to a double bond.

Note: *There is apparently no significant difference in the electronic contribution toward the metal centre as "felt" by the phosphorus atom between three carbonyl ligands on the one hand, and a Cp-ligand on the other.*

A change in the substituent on phosphorus can have a pronounced effect. We are already familiar with an amino substituent's ability to provide a π-donor interaction toward phosphorus, and thus facilitate a significant upfield shift in the phosphorus resonance. Introducing an amino group instead of the Mes* substituent indeed results in a substantial upfield shift of the phosphorus resonance of $\Delta\delta = -250$ to -280 ppm to a value of $\delta_p = 404$ ppm in the example shown in Fig. 7.22. However, if the steric bulk of the amino substituent is increased, π-donor interaction will become impossible, and no upfield shift will be observed. An example is the tetramethyl-piperidine group shown at the right of Fig. 7.22. The steric requirements of the methyl groups prevent the piperidine moiety to align itself coplanarly with the Mo$_2$P three-membered metallacycle. The piperidine is in fact forced into a perpendicular position by the Cp-ligand, resulting in no effective overlap of the nitrogen lone pair with an empty orbital on phosphorus. As a consequence, the phosphorus reso-nance is found at $\delta_p = 648$ ppm for the molybdenum complex and $\delta_p = 593$ ppm for

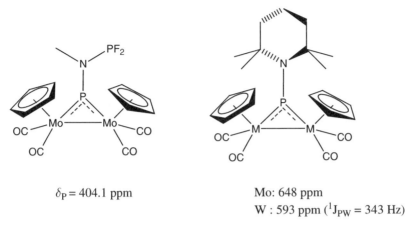

$$\delta_P = 404.1 \text{ ppm}$$

Mo: 648 ppm

W : 593 ppm ($^1J_{PW}$ = 343 Hz)

Fig. 7.22 The dependance of the phosphorus resonance on the availability of a nitrogen lone pair

the homologous tungsten compound. An upfield shift of $\Delta\delta=-55$ ppm from Mo to W is in the expected range.

A further upfield shift of $\Delta\delta=-180$ ppm to a value of $\delta_p=220$ ppm is observed, if the π-donor interaction of the amino substituent on phosphorus is retained, but the three-membered metallacycle is removed (replacement of one of the CpMo(CO)$_2$ groups by a chlorine atom). The P=Mo double bond is now focused onto one molybdenum atom, rather then delocalized over two molybdenum atoms. The phosphorus moiety is a phosphanide rather than a phosphinidene species, and the phosphorus resonance at $\delta_p=220$ ppm is well in keeping with those of terminal phosphanides.

The importance of the π-donor contribution of the phosphorus toward the metal on the ^{31}P-NMR chemical shift can clearly be seen in the two examples depicted in Fig. 7.24. Both complexes have the same core structure, an M$_2$P three-membered metallacycle, and in both complexes the phosphorus atom does not retain a lone pair.

The main difference lies in the electron count of the respective metal fragments. The Fe(CO)$_4$ moiety at the left is a 16 VE fragment requiring only one electron from phosphorus, whereas the Co(CO)$_3$ moiety at the right is a 15 VE fragment requiring two electrons from phosphorus. Therefore, the phosphorus atom has to donate its lone pair in π-interactions toward the two cobalt centers, but forms a σ-donor bond to chromium in the iron complex. Spectroscopically, the missing π-interaction translates into an upfield shift of $\Delta\delta=-369$ ppm.

The two complexes depicted in Fig. 7.25 look deceptively similar to the terminal phosphanide complex seen in Fig. 7.23, but they are indeed terminal phosphinidenes, and the oxidation state of the group 6 metal is $+$IV and thus higher than the oxidation states of the examples in Fig. 7.22, which were $+$II.

The phosphorus resonances of [Cp$_2$Mo=PMes*] are downfield from [{CpMo(CO)$_2$}$_2$(μ-PMes*)] at $\delta_p=687$ ppm, [{Co(CO)$_3$}$_2$(μ-PMes*)] at $\delta_p=664$ ppm,

Fig. 7.23 ³¹P chemical
shift value of a terminal
phosphanide complex

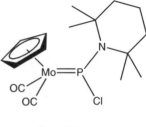

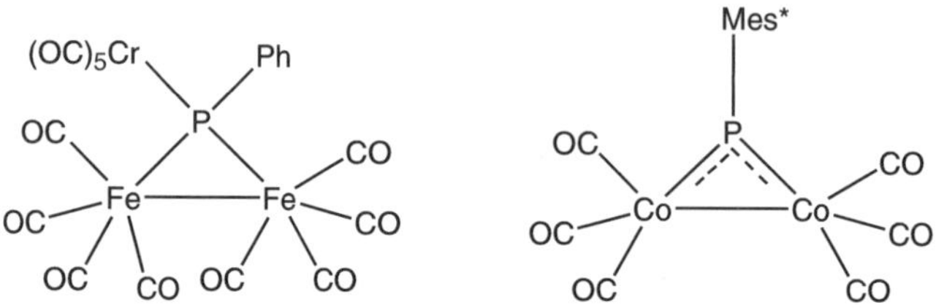

$$\delta_P = 220 \text{ ppm}$$

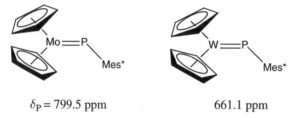

$$\delta_P = 295 \text{ ppm} \qquad\qquad 664 \text{ ppm}$$

Fig. 7.24 Comparison of ³¹P chemical shift values for μ_2- and μ_3-bridging phosphanidenes

$$\delta_P = 799.5 \text{ ppm} \qquad\qquad 661.1 \text{ ppm}$$

Fig. 7.25 The influence of the metal on the ³¹P chemical shift value of terminal phosphinidene complexes

and $[\{CpV(CO)_2\}_2(\mu\text{-PMes}^*)]$ at $\delta_P=657$ ppm. As the strength of the M=P π-interaction is likely the cause of the downfield shift, it is tempting to assume that the metal with the higher oxidation state will be responsible for the greater downfield shift.

The difference between the phosphorus resonances for $[Cp_2Mo=PMes^*]$ and $[Cp_2W=PMes^*]$ is rather large, at $\Delta\delta=138.4$ ppm, further corroborating that the chemical shift depends on the M=P bond order. Shielding of the positive core charge by the full inner d- and f-shells is stronger in tungsten than in molybdenum, and thus the affinity towards the electrons of the M=P π-bond weaker.

The role of the phosphorus lone pair is nicely illustrated in the example shown in Fig. 7.26. The bridging, cationic phosphanide complex on the left has an upfield shifted phosphorus resonance at $\delta_P=-151.3$ ppm (L=CO) and -130.4 ppm (L=PMe$_3$), respectively.

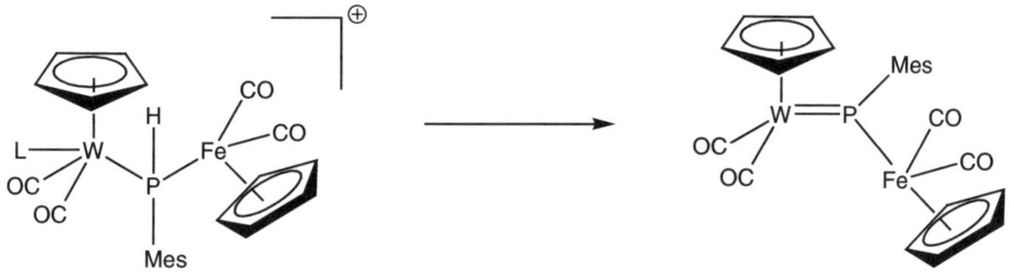

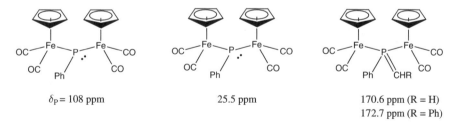

$\delta_P = -151.3$ ppm (L = CO)   352.5 ppm
-130.4 ppm (L = PMe$_3$)

Fig. 7.26 The effect of deprotonation on the ^{31}P chemical shift value of [{CpW(CO)$_3$} (μ-PMes){CpFe(CO)$_2$}]

If the PH proton is abstracted, a neutral bridging phosphinidene species is formed. The newly acquired phosphorus atom lone pair engages in an intramolecular substitution reaction on tungsten, displacing the neutral two electron donor L. We now begin to realize that we were looking at a dimetalla substituted phosphonium cation that resonates $\Delta\delta = -500$ ppm upfield from the bridging phosphinidene complex it turns into upon deprotonation of the phosphorus. Thus, the effect of the π-donor interaction of the phosphorus atom lone pair on the position of the phosphorus resonance is revealed to us in the observed chemical shift differences of $\Delta\delta = 500$ ppm downfield.

Comparison of [{CpW(CO)$_2$}(μ-PMes){CpFe(CO)$_2$}] (see Fig. 7.26) with [{CpMo(CO)$_2$}$_2$(μ-PMes*)] (see Fig. 7.21) shows us that the ring contribution of the three-membered metallacycle appears to be about $\Delta\delta = 300$ ppm, although the π-donor contribution of the phosphorus atom lone pair is not strictly comparable among the two structures (it is focused on tungsten in [{CpW(CO)$_2$}(μ-PMes){CpFe(CO)$_2$}], but shared between both molybdenum atoms in [{CpMo(CO)$_2$}$_2$(μ-PMes*)]). We also have to mention that the choice of metal, Fe/W and Mo/Mo, has an influence on the phosphorus resonance.

As we have just learned, the attempt to synthesize a bridging phosphinidene complex without π-donor contribution from the phosphorus atom lone pair failed for the W/Fe couple due to the ligand lability of the tungsten fragment. The

$\delta_P = 108$ ppm   25.5 ppm   170.6 ppm (R = H)
                                                    172.7 ppm (R = Ph)

Fig. 7.27 Substituent and oxidation effects on the ^{31}P chemical shift values of [{CpFe(CO)$_2$} (μ-PR)]

analogous Fe/Fe complex of a bridging phosphinidene is accessible, however, and the phosphorus resonance is observed at $\delta_p = 108$ ppm (PPh) and 25.5 ppm (PMes), respectively.

Again, notwithstanding the influence of the metal atom (tungsten versus iron), that places the influence of the π-donor interaction of the P=W double bond equivalent to a downfield shift of about $\Delta\delta = 320$ ppm. The electronic equivalence between the metal fragments in [{CpFe(CO)$_2$}$_2$(μ-PMes)] and [{CpW(CO)$_2$} (μ-PMes){CpFe(CO)$_2$}] is given by the P=W π-donor bond for the two electrons missing from tungsten's group number compared to that of iron. The electronic equivalence between [{CpMo(CO)$_2$}$_2$(μ-PMes*)] and [{CpFe(CO)$_2$}$_2$(μ-PMes)] is given by the two electrons from the Mo-Mo bond and the two electrons from the Mo=P π-bond interaction in exchange for the four electrons that the two molybdenum atoms lack compared to the two iron atoms.

Note: *A bridging phosphinidene possesses only one lone pair. It can therefore only contribute two electrons through a π-donor interaction. This "electron shortage" can lead to the formation of metal-metal bonds and thus three-membered metallacycles with a significant ring contribution, resulting in a large downfield shift.*

The molybdenum complex in Fig. 7.28 shows a heteroallene structure with the phosphorus atom in the central position. Electronically, we would expect to see a P=C double bond and a Mo—P covalent bond with the lone pair on phosphorus engaging in a π-donor interaction with the molybdenum atom. This should result in a large downfield shift. However, the linearity of the Mo=P=C heteroallene structure causes considerable shielding, and thus an upfield shift of the phosphorus resonance. It is, therefore, not surprising that the observed phosphorus resonance of $\delta_p = 497$ ppm is somewhat in the middle of [Cp$_2$Mo=P-Mes*] ($\delta_p = 799.5$ ppm) and [CpW(CO)$_2$=P(Mes)-Fe(CO)$_2$Cp] ($\delta_p = 352.5$ ppm), especially as the latter is a tungsten complex.

The difference in the phosphorus resonance for a terminal and a bridging phosphinidene complex can be seen in a series of zirconium complexes shown in Figs. 7.29 and 7.30.

The two terminal zirconium phosphinidene complexes resonate, as expected, at low field, $\delta_p = 465.8$ ppm and $\delta_p = 565.5$ ppm, respectively. The phosphorus resonance in [Cp$^*_2$Zr(H)=P-Mes*]$^-$ is $\Delta\delta = -99.7$ ppm upfield from [Cp$_2$Zr(H)=P-Ph],$^-$ suggesting either a steric influence of the bulky aryl substituent on phosphorus, or an electronic influence of the Cp*-ligand on zirconium.

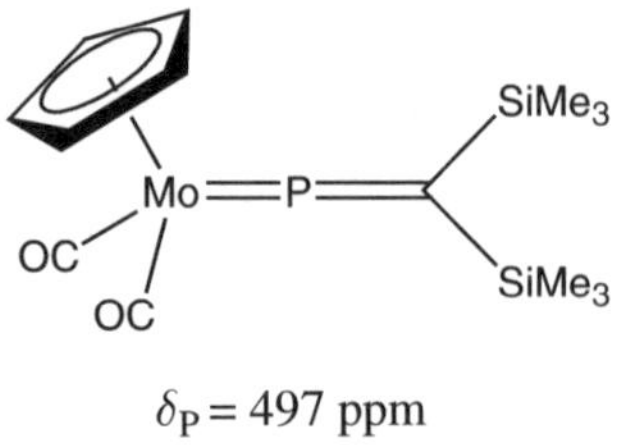

Fig. 7.28 The ^{31}P chemical shift value of [CpMo(CO)$_2$=P=C(SiMe$_3$)$_2$]

$\delta_P = 497$ ppm

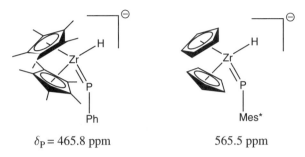

$\delta_P = 465.8$ ppm 565.5 ppm

Fig. 7.29 Dependance of the phosphorus resonance in [Cp$_2$Zr(H)=PR]$^-$ on the substituent R = Ph, Mes*

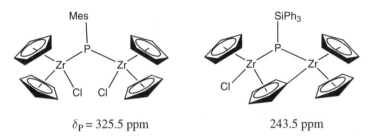

$\delta_P = 325.5$ ppm 243.5 ppm

Fig. 7.30 Bridging phosphinidenes – influence of ring size on the phosphorus resonance

The bridging zirconium phosphinidene complexes in Fig. 7.30 are expected to resonate upfield from the terminal ones in Fig. 7.29, since terminal phosphinidene complexes are far better suited for π-donor interactions of the phosphorus atom towards the metal than are the bridging ligands.

Indeed, the phosphorus resonance in [(Cp$_2$ZrCl)$_2$(μ-PMes)] is found at $\delta_p = 325.5$ ppm or $\Delta\delta = -240$ ppm upfield from [Cp$_2$Zr(H)=P-Ph]$^-$. A further significant upfield shift of $\Delta\delta = -78$ ppm in the phosphorus resonance is observed for [CpZrCl(μ-C$_5$H$_4$) (μ-PSiPh$_3$)ZrCp$_2$], incorporating a four-membered metallacycle (with the bridging C$_5$H$_4$-ring counting as one).

Note: *The anionic zirconium complexes resonate significantly upfield from the neutral, isoelectronic molybdenum and tungsten complexes (Fig. 7.24).*

Note: *The anionic zirconium complexes in Fig. 7.29 resonate significantly upfield from the neutral, isoelectronic zirconium complex [Cp$_2$Zr(PMe$_3$)=P-Mes*] shown in Fig. 7.31.*

The electronic structures of the two complexes depicted in Fig. 7.31 present us with somewhat of a challenge in explaining their respective phosphorus resonances. The chloro-complex on the left clearly has a different electronic contribution from the phosphinidene ligand toward zirconium than the phosphane adduct on the right. A simple electron count on the respective zirconium atoms suggests that in the chloro-complex, the phosphinidene may act as a three electron donor, whereas in

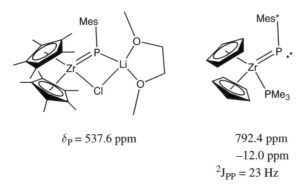

$\delta_P = 537.6$ ppm

792.4 ppm
−12.0 ppm
$^2J_{PP} = 23$ Hz

Fig. 7.31 A puzzle of two zirconium phosphinidene complexes

the phosphane adduct, it acts as a two electron donor only. In the former complex, we would then have a π-donor interaction from phosphorus on top of a single bond, whereas in the latter we see a genuine P=Zr double bond.

In light of this, we would expect the phosphorus resonance of the three electron donor with a π-donor contribution from phosphorus to be downfield from that of the two electron donor species. However, we actually observe the opposite.

As a first step towards understanding this phenomenon, it helps to realize that the structure of the chloro-complex in Fig. 7.31 comes from a x-ray crystal structure determination, and is thus the structure of the complex in the solid state and not in solution. It also helps to note that the zirconium moiety is an anion (Li$^+$ is the counter cation), making the phosphinidene-ligand effectively a two electron donor.

We now look at two complexes with identical electronic structures. In fact, the two complexes are isoelectronic, but the chloro-complex carries a negative charge, and we should expect its phosphorus resonance to be upfield for this reason. The P—Li and Cl—Li donor interactions are very weak, and are not likely to prevail in solution. We can safely neglect them for our discussion of the ^{31}P-NMR chemical shift values.

Note: *Anionic complexes usually resonate upfield from their neutral analogues.*

Figure 7.32 features a base stabilized terminal phosphinidene complex with a phosphorus resonance that is significantly upfield from those previously encountered. According to the x-ray crystal structure determination of this compound, the Fe—P bond order is one, and thus a single bond. It follows that the phosphorus atom has the oxidation state $+I$ and the coordination geometry around phosphorus is tetrahedral with a stereoactive lone pair.

This means that there is no π-bond between phosphorus and iron. From a ^{31}P-NMR spectroscopic viewpoint, we would indeed expect a significant upfield shift of the resonance indicating the absence of a metal-phosphorus double bond.

Note: *The M-P bond order of a phosphinidene complex can easily be determined by ^{31}P-NMR spectroscopy.*

In the compound [TpPFe(CO)$_4$], we have seen the effect of base stabilization on the M-P bond, and thus on the position of the phosphorus resonance. The reaction

Fig. 7.32 The base-stabilized terminal phosphinidene complex [TpPFe(CO)₄]

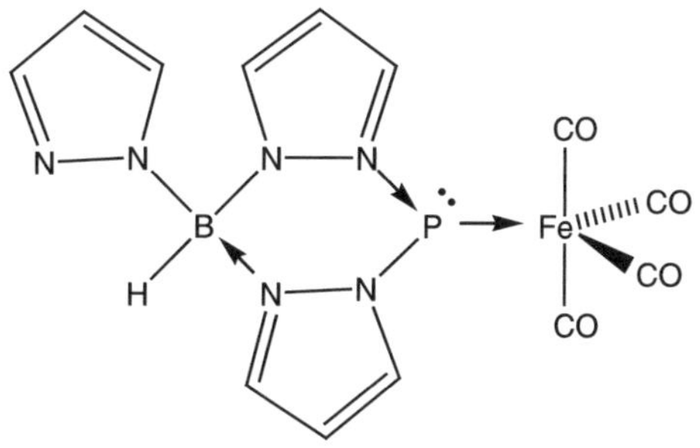

$\delta_P = 281.5$ ppm

was between TpPCl$_2$ (a phosphorus (III) species) and Na$_2$[Fe(CO)$_4$] involving Fe(-II). We concluded that an intramolecular redox reaction takes place, yielding a P(I)/Fe(0) pair. That seems reasonable given the electronegativities of phosphorus and iron.

The phosphorus resonances of the compounds [RPTa(XR')$_3$] in Fig. 7.33 surprise us somewhat, as we would expect the tantalum in these complexes to be in the oxidation state +V rather than +III. The synthesis can be imagined as being between Cl$_2$Ta(XR')$_3$ and Li$_2$PR, and thus between Ta(V) and P(−I). We would expect that given the electronegativities of both phosphorus and tantalum, no intramolecular redox reaction would occur, as that would involve oxidation of phosphorus and reduction of the metal, and therefore the exact opposite to the situation encountered in P(III)/Fe(-II) above.

Let us consider the P(I)/Ta(III) couple nonetheless. The phosphorus atom would retain one of the lone pairs it possesses as a P(I) entity and use the other to form a σ-donor bond to the metal. The empty p-orbital on phosphorus would then be available to accept the π-donor interaction originating from the lone pair on tantalum (M—P backbonding). The result is a trigonal planar phosphorus atom with a stereoactive lone pair and a Ta=P double bond, as observed. However, since the π-donor interaction now involves a filled metal orbital, the phosphorus resonance would be seen relatively upfield.

The other possible explanation for an upfield phosphorus resonance relies on the availability of the lone pairs on oxygen or nitrogen. The tantalum complexes are

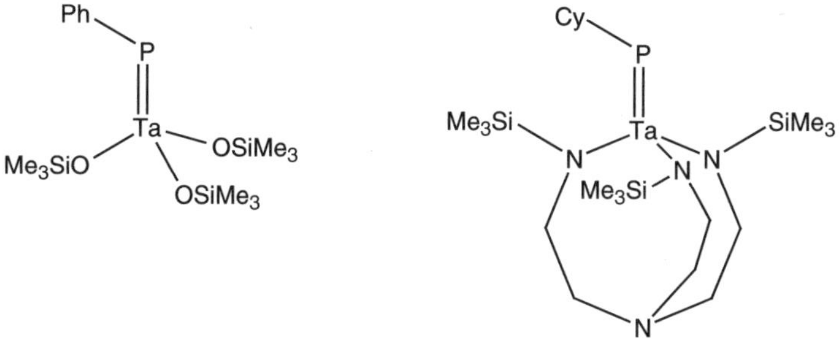

$\delta_P = 334.6$ ppm   209.8 ppm

Fig. 7.33 Comparison of two terminal phosphinidene complexes of tantalum

indeed either siloxides or silylamides, with the tantalum atom bonded to an atom possessing one (nitrogen) or two (oxygen) lone pairs that can, in principle, be used for a π-donor interaction towards the metal. That would in turn shift π-electron density back toward phosphorus, causing an upfield shift in the phosphorus resonance. However, as we have seen earlier (Sect. 6.2), silicon substituents on nitrogen (oxygen) effectively inactivate its lone pair(s) and make it unavailable to the metal. We have, therefore, no other option but to explain the significant upfield shift of the phosphorus resonance with the P(I)/Ta(III) couple.

7.5 Phosphaacetylene Complexes

There are two principal coordination modes for a phosphaacetylene coordinating to a metal, end-on and side-on (see Fig. 7.36). In the side-on coordination, the bonding can either be described as a π-donor interaction of the triple bond, or as two single bonds (M-C and M-P). These three bonding modes should be distinguishable by their phosphorus resonances.

The end-on coordination where the phosphorus lone pair forms a σ-donor interaction to the metal should result in the accustomed small downfield shift.

If we look at Fig. 7.34, we find that the phosphorus resonance for the molybdenum complex is at $\delta_P = 57.0$ ppm, and thus $\Delta\delta = 22.6$ ppm downfield from that of the free ligand. Interestingly, the chemical shift of the tungsten complex is actually $\Delta\delta = -10.0$ ppm upfield.

This finding is corroborated by two other complexes shown in Fig. 7.35, an anionic iron (0) and a molybdenum (0) complex. The phosphorus resonances of these complexes are observed at $\delta_P = -13$ ppm and $\delta_P = -7.2$ ppm, respectively, or $\Delta\delta = 56.2$ ppm and $\Delta\delta = 62.0$ ppm downfield from the free ligand. This is in line with the behavior of phosphane ligands coordinating to transition metals whose phosphorus resonances likewise experience a downfield shift of similar magnitude (see Sect. 7.1).

The same trend can be seen in the molybdenum(0) complexes depicted in Table 7.14 carrying two phosphaacetylene ligands *trans* to each other. The downfield shift is somewhat larger than in the similar complexes shown in Fig. 7.36. That can be expected, since the phosphaacetylene ligand is a better π-acceptor than either dinitrogen or hydride.

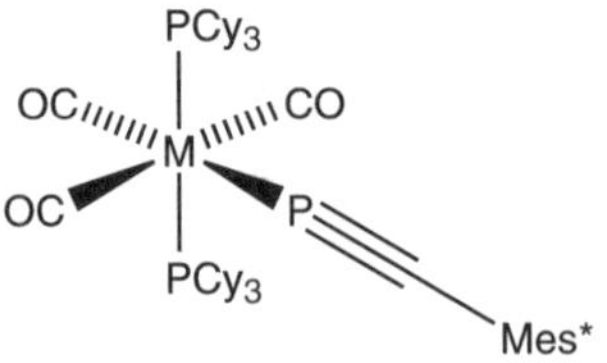

Fig. 7.34 The ^{31}P chemical shift values for $[M(CO)_3(PCy_3)_2P\equiv CMes^*]$

M =	Mo	W
δ_P =	57.0 ppm	24.4 ppm

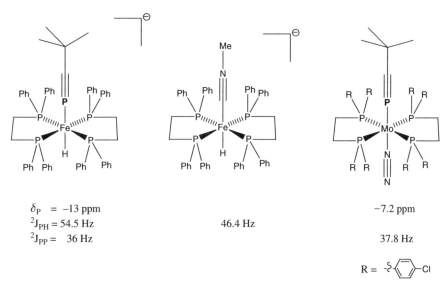

δ_P = –13 ppm
$^2J_{PH}$ = 54.5 Hz
$^2J_{PP}$ = 36 Hz

46.4 Hz

–7.2 ppm

37.8 Hz

$R = $ –ξ⟨◯⟩–Cl

Fig. 7.35 Some phosphaalkyne complexes of iron and molybdenum

Figure 7.36 illustrates the two side-on bonding modes of a phosphaacetylene to a transition metal fragment. The problem is similar to that of an olefin or acetylene bonding side-on to a metal, and amounts to the question of whether a π- or a σ-complex is formed. In a π-complex, the multiple bond serves as a two electron donor towards the metal, whereas in a σ-complex, the bonding interaction is seen as two single bonds between the metal and the two carbon atoms. It is not easy to settle the argument for the olefin and acetylene complexes as spectroscopic and structural evidence are largely indecisive.

Table 7.14 Modifying the ^{31}P chemical shift values of molybdenum phosphaalkyne complexes

	M	R	R'	δ_p [ppm]	J_{PP} [Hz]
	Mo	Bu^t	Et	10.0	40.3
	Mo	Bu^t	Ph	0.1	38.4
	Mo	Bu^t	p-Tol	0.3	37.8
	Mo	Ad	Et	7.7	40.3
	W	Bu^t	Ph	–16.9	26.9

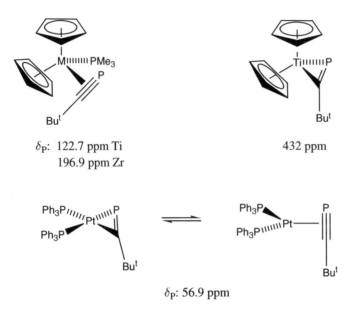

δ_P: 122.7 ppm Ti
      196.9 ppm Zr

432 ppm

δ_P: 56.9 ppm

Fig. 7.36 The possible bonding modes of side-on phosphaalkyne complexes of selected transition metals

For a phosphaacetylene complex, however, this is different, as ^{31}P-NMR spectroscopy is a powerful tool to detect metallacycles as well as probing for the extent of π-bonding interactions. A first assumption would be that a π-complex would not contain a metallacycle, and thus would not incur a downfield ring contribution, whereas the σ-complex would. In addition, the π-complex would essentially retain its phosphaacetylene character, but the σ-complex would turn into a phosphaolefin entity, which is again significantly shifted downfield from a phosphaacetylene.

Bearing this in mind, we examine the two structures shown in Fig. 7.36 of group 4 phosphaacetylene adducts. In both complexes, the phosphaacetylene is bonded side-on, but in the complex to the left, the metal carries a phosphane ligand that is absent in the complex to the right. The effect on the phosphorus resonance is remarkable. They are found at $\delta_p = 127.7$ ppm (Ti), 196.9 ppm (Zr), and 432 ppm, respectively. The phosphorus resonance at $\delta_p = 432$ ppm compared to $\delta_p = 127.7$ ppm shows us that we look at a σ- and a π-complex, respectively, separated by a downfield shift of approximately $\Delta\delta = 304$ ppm.

However, the trend in the phosphorus resonances of the isostructural titanium and zirconium complexes strikes us as odd. We know that the chemical shift moves upfield, not downfield, as one moves down the triad. In our two examples, the trend is reversed. The only logical explanation would seem that the larger zirconium atom favours the σ-complex over the π-complex, and thus the downfield shift seen in the transition from titanium to zirconium reflects the extent of the π-complex present in the true structure of this complex.

Unfortunately, this trend is not universal for side-on bonded phosphaacetylene complexes, as can be clearly seen from the platinum example in Fig. 7.36. The phosphorus resonance is observed at $\delta_P = 56.9$ ppm, indicating a π-complex.

The reasons for this behavior can only be guessed at, and might include the relative electron richness of the metal and its geometrical preference (tetrahedral for group 4, and planar 16 VE for platinum).

Note: *A phosphaacetylene side-on σ-complex resonates significantly downfield from the respective π-complex.*

7.6 Agostic M—P—H Interactions

Phosphanes and phosphanides carrying a hydrogen substituent can, in principle, bond to a transition metal in two ways, either through phosphorus alone or by phosphorus and hydrogen. This latter bonding mode, where the hydrogen atom effectively bridges the phosphorus and transition metal atoms, is known as an agostic P-H--M interaction. This bonding interaction results in something like a three-membered MPH ring that apparently **does not** experience a downfield shift usually observable in three-membered phosphametallacycles.

Quite to the contrary, this agostic interaction is distinguished by a considerable upfield shift in the phosphorus resonance, and a coupling constant that is approximately half the value of an ordinary $^1J_{PH}$ coupling constant. A good example is a series of [(CpMoCO)$_2$(μ-PRR')(μ-PHRR')]$^+$ complexes (see Table 7.15). Here, there are two phosphanide bridges, one with and one without a hydrogen substituent. The resonance of the one with the additional hydrogen substituent is expected to appear downfield from the other, since protonation of the phosphorus atom should result in a decrease of electron density on this phosphorus atom, and thus deshielding. However, the opposite is observed. The resonance moves upfield as apparently the affected phosphorus atom gains more from the additional M-P backbonding than it loses to the hydrogen through protonation. In addition, the $^1J_{HP}$ coupling constant drops to 100–130 Hz, approximately half the value of 200–300 Hz usually associated with a $^1J_{HP}$ coupling constant.

Table 7.15 Agostic P-H--M interactions in a series of [(CpMoCO)$_2$(μ-PRR')(μ-PHRR')]$^+$ complexes

	R	R'	δ_P (μ)	δ_P (HP)
	Et	Et	145.0	91.4 (103 Hz)
	Cy	Cy	161.4	89.7 (134 Hz)
			163.2	90.2 (127 Hz)
	Cy	Ph	145.5	100.4 (132 Hz)
			141.1	103.5 (132 Hz)

Another example is a series of platinum complexes [{Pt(PR$_2$R")}$_2$(μ-PHR$_2$) (μ-PR'$_2$)] (see Fig. 7.37), where the same phenomena are observed as in the molybdenum complexes. The phosphorus atom engaged in the P-H--M agostic interaction is shielded compared to the other phosphorus atom, and the recorded ^{1}J$_{PH}$ coupling constants are much reduced to 136–151 Hz. The two terminal phosphanes have different chemical shifts, as expected, with the phosphane bonded to the five coordinate platinum featuring a phosphorus resonance upfield to the phosphane bonded to the four coordinate platinum.

A rather intriguing example is presented in Fig. 7.38. Here we look not only at the effect of the agostic P-H--M interaction that again results in the now familiar chemical shift differences of the two phosphanido bridges, but also at the effect of protonation. In the neutral symmetrical complex, only one phosphorus resonance is observed at $\delta_p = 292.7$ ppm. In the protonated complexes, the expected two resonances are recorded at $\delta_p = 402.2$ ppm and at $\delta_p = 303.1$ ppm, both downfield from the signal in the unprotonated parent compound. The overall downfield shift is, of course, the result of the protonation of the complex, whereas the agostic P-H--M interaction then moves the appropriate phosphorus resonance back upfield by $\Delta\delta \approx -100$ ppm (from $\delta = 402.2$ ppm to $\delta = 303.1$ ppm).

That the characteristic upfield shift of this agostic MPH-entity can be used as a diagnostic tool is evident in the complex [Cp$^*_2$Zr(η^3-PHPhPPhPPh)], featuring

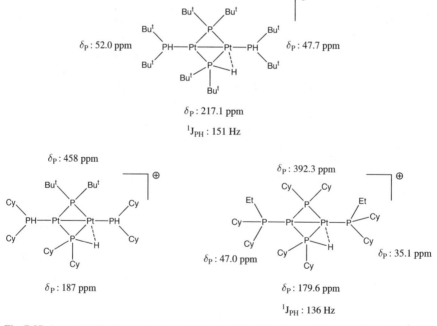

Fig. 7.37 Agostic P-H--M interactions in a series of [{Pt(PR$_2$R")}$_2$(μ-PHR$_2$)(μ-PR'$_2$)] complexes

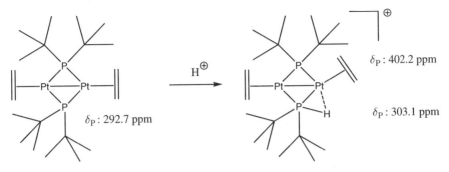

Fig. 7.38 Effect of agostic P-H--M interactions on the phosphorus chemical shift in some platinum ethylene complexes with bridging phosphanide ligands

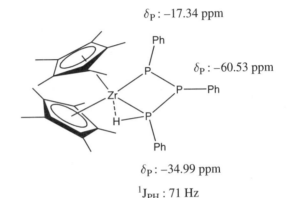

Fig. 7.39 Effect of agostic P-H--M interactions on the phosphorus chemical shift in [Cp*$_2$Zr(η^3-PHPhPPhPPh)]

three references in the ^{31}P-NMR spectrum at $\delta_p = -17.34$ ppm, $\delta_p = -34.99$ ppm ($^1J_{PH} = 71$ Hz), and $\delta_p = -60.53$ ppm, respectively. The small $^1J_{PH}$ coupling constant of 71 Hz, and the two inequivalent chemical shift values of $\delta_p = -34.99$ ppm and $\delta_p = -17.34$ ppm, are indicative of an agostic P-H--M interaction. A terminal Zr-H bond would not be consistent with the observed chemical shifts and coupling constants, and neither would be a terminal P—H bond.

Note: *The characteristic upfield shift and reduced $^1J_{PH}$ coupling constants associated with agostic P-H--M interactions can be used to assign the ^{31}P-NMR spectrum correctly and help to determine the structure of the compound spectroscopically.*

7.7 Naked Phosphorus

An interesting array of phosphorus ligands is realized by species where the phosphorus atoms are bonded only to metal or phosphorus atoms. They carry no substituents and are called "naked phosphorus". Their chemical shift values cover a very broad range from $\delta_P = -488$ ppm for white phosphorus P_4 to at least $\delta_P = 959$ ppm for $[(Cp^*_2Zr)_2(\mu\text{-}P_2)]$ ($Cp^* = C_5Me_5$), and depend both on the electronic properties of the P_n-species and the particular bonding interactions to the metal atoms.

The upfield end of the range is marked by white phosphorus P_4 itself that can coordinate to a transition metal either by a vertex atom or with both phosphorus atoms of a common edge bonding to one or two transition metal atoms. It is important to realise that one or more phosphorus atoms in the P_4 tetrahedron can be replaced by an isolobal metal fragment (see Fig. 7.40), with only a moderate downfield shift (for a summary of the isolobal theory see the box story following Chap. 7).

Other naked phosphorus units include the "aromatic" species P_4^{2-}, P_5^- and P_6^{4-} that are not stable as free ligands H_2P_4, HP_5 and H_4P_6. Alkali salts of P_4^{2-} and P_6^{4-} can be isolated, whereas MP_5 ($M = Na$, K) is only stable in dilute solutions or in form of transition metal complexes where the P_5^- ring takes the place of a cyclopentadienide moiety. The closely related species P_7^{3-} and P_{11}^{3-} will not be discussed here, as they exhibit a complicated coupling pattern that cannot be discussed from a first-order point of view.

We will complete our list of naked phosphorus units with P^{3-}, P_2^{2-}, P_4^{4-} and P_3^{5-} (see Fig. 7.41), all of which can be generated in the coordination sphere of transition metals, either by the elimination of subsituents on phosphorus, or by transfer of phosphorus units from one transition metal to the other.

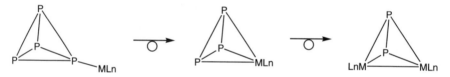

Fig. 7.40 Isolobal relationships in M_nP_{4-n} tetrahedra

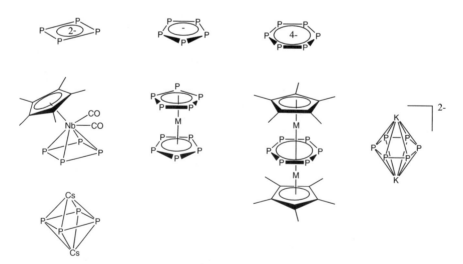

Fig. 7.41 Aromatic naked phosphorus ligands and their complexes

The ^{31}P-NMR resonance of white phosphorus P$_4$ sits on the far upfield end of the ^{31}P chemical shift range. There are various reports for its chemical shift value, essentially ranging from $\delta_P = -527$ to -488 ppm. Most interestingly, the same group reports the chemical shift of white phosphorus as $\delta_P = -522$ and $\delta_P = -527$ ppm in the same solvent (CD$_2$Cl$_2$) leaving one to speculate that the chemical shift might depend on the degree of dryness of white phosphorus (usually stored under water). Coordination of P$_4$ onto a transition metal should result in a small downfield shift of the resonance in keeping with that of the phosphane ligands.

Coordination of the P$_4$ tetrahedron to transition metals can happen in η^1-, η^2- and μ^2,η^2-fashion, resulting in characteristic chemical shift values and coupling patterns unless the P$_4$ tetrahedron shows fluxional behavior in its bonding, something that is frequently observed. The η^1-coordination is observed in *mer*-[W(CO)$_3$(PCy$_3$)$_2$ (η^1-P$_4$)] (toluene-d8) with an A$_2$MX$_3$ spin system overall and an AX$_3$ system for the P$_4$ tetrahedron. The phosphorus atom bonded to tungsten is observed as a quartet at $\delta_P = -422$ ppm ($^1J_{PP} = 204$ Hz) and the other three phosphorus atoms at $\delta_P = -473$ ppm (d, $^1J_{PP} = 204$ Hz). Coordination to tungsten results in a $\Delta\delta = 51$ ppm downfield shift relative to the uncoordinated phosphorus atoms of the coordinated P$_4$ unit. We can safely assume that these three phosphorus atoms are themselves shifted downfield by about $\Delta\delta = 10$ ppm, similar to a monocoordinated diphosphane. Similar chemical shifts are observed for [Re(CO)$_2$(triphos)(η^1-P$_4$)] with $\delta_P = -391$ ppm and $\delta_P = -489$ ppm, respectively (see Fig. 7.42 and Table 7.16).

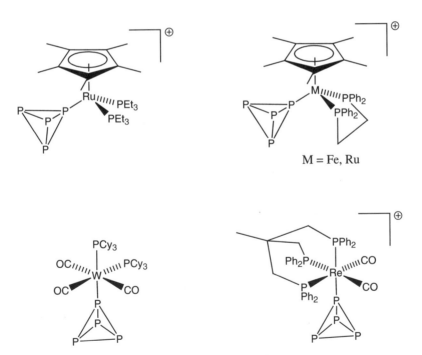

M = Fe, Ru

Fig. 7.42 Examples for η^1-P$_4$ complexes of transition metals

Table 7.16 ^{31}P-NMR chemical shift values for transition metal P_4-complexes

Complex	δ_p (η) [ppm]	δ_p [ppm]	$^1J_{pp}$ [Hz]
[Cp*Ru(PEt$_3$)$_2$(η^1-P$_4$)] BPh$_4$	−332.05 (t)	−480.96 (d)	228.9
[Cp*Ru(dppe)(η^1-P$_4$)] BPh$_4$	−308.46 (t)	−490.29 (d)	233.5
[Cp*Fe(dppe)(η^1-P$_4$)] BPh$_4$	−299.54 (t)	−482.12 (d)	228.9
[Re(CO)$_2$(triphos)(η^1-P$_4$)]$^+$	−391	−489	
[W(CO)$_3$(PCy$_3$)$_2$(η^1-P$_4$)]	−422 (t)	−473 (d)	204

In the series [Cp*Ru(PEt$_3$)$_2$(η^1-P$_4$)]BPh$_4$ and [Cp*M(dppe)(η^1-P$_4$)]BPh$_4$ (M = Fe, Ru) the usual trends concerning the metal and the substituents on the coligands are observed (see Fig. 7.42 and Table 7.16). The signals in the iron complex are shifted downfield from the signals in the respective ruthenium complex, whereas the PEt$_3$ ligand, being the weaker π-acceptor compared to dppe, causes an upfield shift in the pair of ruthenium complexes.

When the P$_4$ unit coordinates η^2 to the same metal atom, a three-membered ring is formed, resulting in a significant downfield shift. An example is [Rh(η^2-P$_4$)(PPh$_3$)$_2$] where the P$_4$ unit shows two multiplets in the ^{31}P-NMR spectrum at $\delta_p = -279.4$ and $\delta_p = -284.0$ ppm, respectively, about $\Delta\delta = 200$ ppm downfield from uncoordinated P$_4$, and approximately $\Delta\delta = 100$–150 ppm downfield from η^1-P$_4$. It is interesting to note that both phosphorus nuclei, the two coordinated to Rh(I) and the two that are not, have approximately the same chemical shift value. The pattern is that of an A$_2$B$_2$ spin system. The P$_4$ unit is evidently affected as a whole, irrespective of which individual phosphorus atom is bonded to the metal. Unfortunately, no variable temperature (VT) NMR studies were reported.

The Fe(0) complex [{Fe(CO)$_4$}$_3$P$_4$] shows only one ^{31}P-NMR signal as a singlet at $\delta_p = 21$ ppm (see Fig. 7.43). Its x-ray crystal structure, however, shows two Fe(CO)$_4$ units bridging one of the P-P edges each, such that each phosphorus atom is bonded to iron. The third Fe(CO)$_4$ unit is coordinated terminally to one of the four phosphorus atoms at the vertices (see Fig. 7.43). Such a structure can only be reconciled with the ^{31}P-NMR spectrum in solution, if fluxional behavior is assumed. The significant downfield shift between the isoelectronic Rh(I) and Fe(0) complexes is due to the coordination of an additional metal fragment on each of the four vertices of the P$_4$ unit on the NMR time scale.

Not only can the P$_4$ unit be coordinated terminally or bridging onto transition metals, but one or more of the phosphorus atoms in the P$_4$ unit can be substituted by an isolobal transition metal fragment without destroying the cluster (see Fig. 7.44). Examples for monosubstitution of phosphorus by a 15 VE metal fragment can be

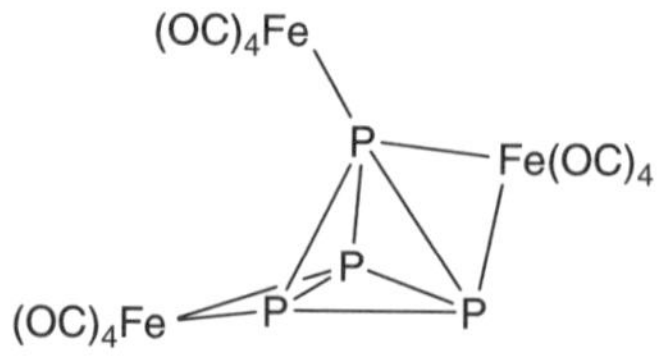

Fig. 7.43 Solid state structure of [{Fe(CO)$_4$}$_3$P$_4$]

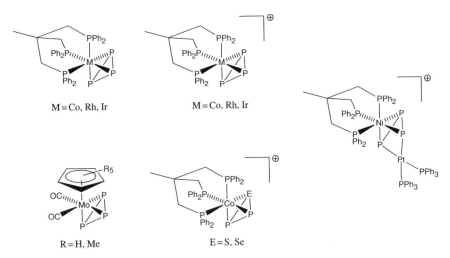

Fig. 7.44 Examples for MP$_3$-complexes

found in Table 7.17. Incorporation of a transition metal into the P$_4$ unit results in a similar downfield shift as coordination along an edge of the unsubstituted P$_4$ unit.

The chemical shifts listed in Table 7.17 show the usual trends, namely an upfield shift as one moves down a triad (Co – Rh – Ir) and a downfield shift upon introduction of a positive charge (Co – Ni). It is interesting that in these series, the second member of a triad (Rh, Pd) is always slightly out of tune. Coordination of a second transition metal onto the MP$_3$ tetrahedron as in [(triphos)Ni({P$_3$Pt(PPh$_3$)$_2$}]BF$_4$ again results in a significant downfield shift ($\Delta\delta$ = 52.2 ppm) of the phosphorus resonance.

Substitution of a second phosphorus atom by a 15 VE transition metal fragment yields tetrahedrons of the general formula M$_2$P$_2$. The introduction of a second transition metal results in the expected downfield shift. The phosphorus resonance can now be found around δ_P = −50 ppm, as can be seen in the molybdenum complexes depicted in Fig. 7.45. Coordination of a [Cr(CO)$_5$] fragment

Table 7.17 ^{31}P-NMR chemical shift values for MP$_3$-complexes

Complex	Solvent	δ_P [ppm]
[CpMo(CO)$_2$(η^3-P$_3$)]	C$_6$D$_6$	−351.5
[Cp*Mo(CO)$_2$(η^3-P$_3$)]	C$_6$D$_6$	−336.5
[(triphos)Co(η^3-P$_3$)]	CD$_2$Cl$_2$	−276.2
[(triphos)Rh(η^3-P$_3$)]	CD$_2$Cl$_2$	−270.8
[(triphos)Ir(η^3-P$_3$)]	CD$_2$Cl$_2$	−312.9
[(triphos)Ni(η^3-P$_3$)]BF$_4$	CD$_2$Cl$_2$	−155.7
[(triphos)Pd(η^3-P$_3$)]BF$_4$	CD$_2$Cl$_2$	−132.9
[(triphos)Pt(η^3-P$_3$)]BF$_4$	CD$_2$Cl$_2$	−217.4
[(triphos)Co(η^3-P$_2$S)]BF$_4$	CD$_2$Cl$_2$	−210.9
[(triphos)Co(η^3-P$_2$Se)]BF$_4$	CD$_2$Cl$_2$	−145.5
[(triphos)Ni({P$_3$Pt(PPh$_3$)$_2$}]BF$_4$	CD$_2$Cl$_2$	−103.5

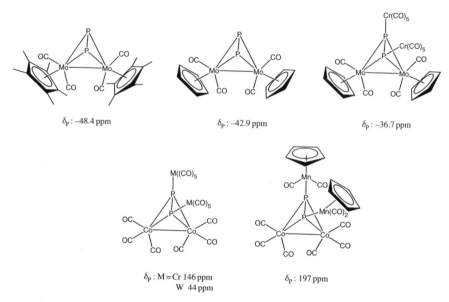

δ_P: −48.4 ppm δ_P: −42.9 ppm δ_P: −36.7 ppm

δ_P: M = Cr 146 ppm δ_P: 197 ppm
W   44 ppm

Fig. 7.45 Examples for M_2P_2-complexes

to the phosphorus lone pairs results only in a small downfield shift of about $\Delta\delta = 6$ ppm, in contrast to a series of cobalt complexes with identical geometry whose phosphorus resonances are observed between $\delta_p = 44$ and $\delta_p = 197$ ppm. In the sequence W(0), Cr(0), Mn(I) with phosphorus resonances of $\delta_p = 44$, 146, and 197 ppm, respectively, the familiar trends are again mirrored (downfield shift for the transition W to Cr and for the transition of the oxidation state from 0 to $+I$).

The M_2P_2 tetrahedron can be extended by either an additional M_2 group or a second P_2 unit. The result is a pair of edge-fused tetrahedra. Examples are $[\{(CpMo(CO)_2)_2 (\mu,\eta^2-P_2)\}\{Re_2Br_2(CO)_6\}]$ and $[Cp^*Mo(CO)(\mu,\eta^2-P_2)_2(CO)MoCp^*]$ (see Fig. 7.46), whose phosphorus resonances are upfield from the unfused M_2P_2 tetrahedron.

If a transition metal fragment with more than 15 VE is used, the M_2P_2 tetrahedron is broken up. Due to the additional electrons of the transition metal fragment, either the M-M bond or the P-P bond becomes obsolete. The result is a downfield

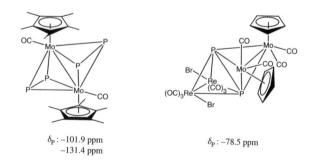

Fig. 7.46 Structures of edge-fused Mo_2P_4 and $Re_2P_2Mo_2$-complexes

δ_p: −101.9 ppm
     −131.4 ppm

δ_p: −78.5 ppm

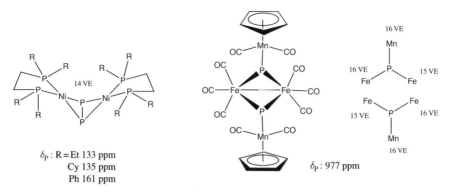

δ_P : R = Et 133 ppm
         Cy 135 ppm
         Ph 161 ppm

δ_P : 977 ppm

Fig. 7.47 Structures of M_2P_2-complexes with metal fragments not isolobal to P

shift, as can be seen in the examples shown in Fig. 7.47. Please note that the 14 VE nickel fragment has square planar geometry, making it equivalent to a 16 VE metal fragment.

The dramatic downfield shift observed in the phosphorus resonances between $[(dppeNi)_2(\mu,\eta^2-P_2)]$ and $[Fe_2(CO)_6(\mu_3-P)\{MnCp(CO)_2\}_2]$ of about $\Delta\delta=800$ ppm is, of course, not a result of the opening of the P-P bond or the coordination of another transition metal using the phosphorus lone pair, but is caused by a change of geometry on the phosphorus atom. Up until now, we have encountered the phosphorus atom almost always embedded in a tetrahedral environment. Now, the phosphorus atom is forced to adopt a trigonal planar geometry. In $[Fe_2(CO)_6(\mu_3-P)\{MnCp(CO)_2\}_2]$, the phosphorus atoms utilise all five electrons to bond to the three transition metals. The trigonal planar geometry is adopted in a desperate attempt to salvage some electron density by metal to phosphorus backbonding, for which the phosphorus atom has to place the empty p-orbital perpendicular to the three σ-bonds. The low electron density on phosphorus is responsible for the large downfield shift of $\delta_P=977$ ppm. That this is indeed the case and not the formation of a three-membered metallacycle is already evident from the $[(dppeNi)_2(\mu,\eta^2-P_2)]$ complex, but can also be seen from comparison with $[CpW(CO)_2\{Cr(CO)_5\}_2(\mu_3-P)]$ and $[\{Cr(CO)_5\}_2(\mu-PBu^t)]$. The latter does not feature a three-membered ring, but a trigonal planar phosphorus atom and a chemical shift of $\delta_P=1362$ ppm, a full $\Delta\delta=385$ ppm downfield from $[Fe_2(CO)_6(\mu_3-P)\{MnCp(CO)_2\}_2]$.

An example for a M_3P tetrahedron is available with M = Zr (see Fig. 7.48). In $[\{CpZr(\mu-C_5H_4)\}_3(\mu_3-P)]$, the phosphorus atom is a three electron donor and has tetrahedral geometry, instead of being a five electron donor with trigonal planar geometry as in $[Fe_2(CO)_6(\mu_3-P)\{MnCp(CO)_2\}_2]$. As a consequence, the chemical shift is found at $\delta_P=782.6$ ppm, considerably upfield from the value of $\delta_P=977$ ppm for the Fe$_2$Mn complex. However, a chemical shift of $\delta_P=783$ ppm is still very much deshielded. In the zirconium complex, the phosphorus atom is bonded to three electron-deficient early transition metals with d^0-configuration, resulting in a low electron density on phosphorus.

Aside from the P_4 theme with the series of isolobal M_nP_{4-n} tetrahedra, the zirconium chemistry affords an interesting case study in the principles that govern

Fig. 7.48 Highly deshielded μ₃-P transition metal complexes

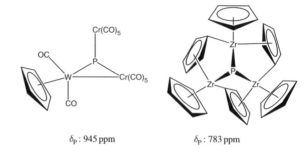

δ_P : 945 ppm δ_P : 783 ppm

the chemical shifts in ³¹P-NMR spectroscopy. We shall look at the compounds [Cp*₂Zr(η²-P₂)], [{Cp*₂Zr}₂(μ,η²-P₂)], and [Cp*₂Zr(η²-P₃)] (see Fig. 7.49).

The three complexes distinguish themselves from the tetrahedral M_nP_{4-n} structures we have discussed so far by their very significantly deshielded phosphorus resonances. In [Cp*₂Zr(η²-P₃)], we still observe the familiar MP₃ motive, but with very unfamiliar shift patterns. We now note a doublet at $\delta_P = 490.4$ ppm, and an accompanying triplet at $\delta_P = 245.6$ ppm. Not only are the two signals some $\Delta\delta = 400$–650 ppm downfield from the singlet for the NiP₃⁺-complexes, but they have also become inequivalent. Evidently, the zirconium atom is less strongly bonded to the central phosphorus atom than to the other two, an assessment borne out by the crystal structure. The Cp*₂Zr fragment is a 14 VE unit, and thus does not fit into the 15 VE pattern of isolobal replacements in the M_nP_{4-n} series. As a consequence, we observe a significant downfield shift similar to the one observed for [(dppeNi)₂(μ, η²-P₂)]. The larger extent of the downfield shift is explained by the electronic differences in zirconium (an early transition metal) and nickel (a late transition metal). Please note

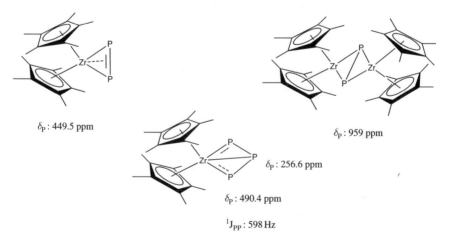

δ_P : 449.5 ppm δ_P : 959 ppm

δ_P : 256.6 ppm

δ_P : 490.4 ppm

¹J_{PP} : 598 Hz

Fig. 7.49 Structures of zirconium complexes of P₂ and P₃ ligands

that every phosphorus atom in the structure is part of at least one three-membered metallacycle, but the magnitude of the downfield shift is strongly dependent on the Zr-P bond length and the formal charge on phosphorus.

In $[Cp^*_2Zr(\eta^2\text{-}P_2)]$, the phosphorus signal is likewise downfield shifted to $\delta_p = 449.5$ ppm, a value consistent with a ZrP_2 three-membered metallacycle. Adding another Cp^*_2Zr-fragment onto $[Cp^*_2Zr(\eta^2\text{-}P_2)]$ results in an additional downfield shift of $\Delta\delta = 409.5$ ppm to $\delta_p = 959$ ppm in $[\{Cp^*_2Zr\}_2(\mu,\eta^2\text{-}P_2)]$, as expected. The central P_2-unit is now part of two ZrP_2 three-membered metallacycles, each of them involving an electron deficient 14 VE early transition metal fragment. As involvement in multiple metallacycles results in additive chemical shift increments, this dramatic downfield shift in going from $[Cp^*_2Zr(\eta^2\text{-}P_2)]$ to $[\{Cp^*_2Zr\}_2(\mu,\eta^2\text{-}P_2)]$ is hardly surprising.

Box Story: Isolobal Theory

The isolobal theory, developed by R Hoffman (an excellent account was given by him in the acceptance speech for the Nobel Prize; see bibliography), to compare ML_n cluster fragments from transition metal clusters with CH_l and BH_m cluster fragments from carboranes and boranes has become a very powerful tool to understand the properties and reactivities of main group and transition metal moieties.

It is often found that a transition metal fragment containing ten electrons more than a corresponding main group fragment has similar properties. This has been attributed to the 18 VE rule. However, a simple electron count is not enough to establish an isolobal relationship. For an isolobal relationship to be established, the molecular orbitals used to bind to other fragments (known as frontier orbitals) need to display a close similarity.

Definition "Two fragments are isolobal if the number, symmetry properties, approximate energy and shape of their frontier orbitals, and the number of electrons occupying them are similar – not identical, but similar." [R. Hoffmann 1982].

Examples for an isolobal relationship (P is isolobal to $CpMo(CO)_2$) and for the lack of an isolobal relationship (P is not isolobal to Nidppe) are provided in Figs. 7.50 and 7.51. It is quite obvious that the two frontier orbitals of the nickel fragment are not equivalent to the three frontier orbitals of the molybdenum fragment. The consequence of the present case, of course, is that substitution of P with $CpMo(CO)_2$ would result in retention of all of the bonds in the P_4 tetrahedron, whereas substitution with Nidppe would result in a different bonding pattern, as a bond is "missing". In fact, $[(dppeNi)_2(\mu\text{-}P_2)]$ does not possess a Ni—Ni bond, whereas $[\{CpMo(CO)_2\}_2(\mu\text{-}P_2)]$ does possess a Mo—Mo bond.

Box Story: (continued)

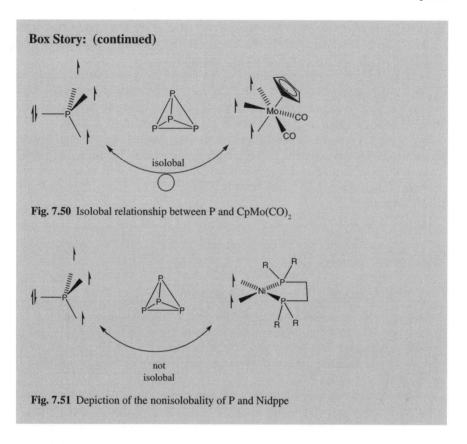

Fig. 7.50 Isolobal relationship between P and CpMo(CO)₂

Fig. 7.51 Depiction of the nonisolobality of P and Nidppe

Bibliography

Alvarez C M, Galan B, Garcia M E, Riera V, Ruiz M A, Vaissermann J, Organometallics 22 (2003) 5504.

Alvarez C M, Garcia M E, Martinez M E, Ramos A, Ruiz M A, Saez D, Vaissermann J, Inorg Chem 45 (2006) 6965.

Alvarez C M, Garcia M E, Ruiz M A, Connelly N G, Organometallics 23 (2004) 4750.

Alvarez M A, Anaya Y, Garcia M E, Riera V, Ruiz M A, Organometallics 23 (2004) 433.

Alvarez M A, Garcia M E, Martinez M E, Ramos A, Ruiz M A, Saez D, Vaissermann J, Inorg Chem 45 (2006) 6965.

Arif A M, Cowley A H, Norman N C, Orpen A G, Pakulski M, Organometallics 7 (1988) 309.

Arliguie T, Ephritikhine M, Lance M, Nierlich M, J Organomet Chem 524 (1996) 293.

Bartsch R, Hietkamp S, Morton S, Stelzer O, J Organomet Chem 222 (1981) 263.

Bender R, Bouaoud S-E, Braunstein P, Dusausoy Y, Merabet N, Raya J, Rouag D, J Chem. Soc, Dalton Trans (1999) 735.

Bender R, Braunstein P, Dedieu A, Ellis P D, Huggins B, Harvey P, Sappa E, Tripicchio A, Inorg Chem 35 (1996) 1223.

Binger P, Biedenbach B, Herrmann A T, Langhauser F, Betz P, Goddard R, Krüger C, Chem Ber 122 (1990) 1617.

Bonanno J B, Wolczanski P T, Lobkovsky E B, J Am Chem Soc 116 (1994) 1159.

Brunner H, Rötzer M, J Organomet Chem 425 (1992) 119.

Burckett-St. Laurent J C T R, Hitchcock P B, Kroto H W, Nixon J F, J Chem Soc, Chem Comm (1981) 1141.

Cowley A H, Barron A R, Acc Chem Res 21 (1988) 81.

Cowley A H, Geerts R L, Nunn C M, J Am Chem Soc 109 (1987) 6523.

Cowley A H, Norman N C, Quashie S, J Am Chem Soc 106 (1984) 5007.

Cowley A H, Pellerin B, J Am Chem Soc 112 (1990) 6734.

de los Rios I, Hamon J-R, Hamon P, Lapinte C, Toupet L, Romerosa A, Peruzzini M, Angew Chem Int Ed 40 (2001) 3910.

di Vaira M, Stoppioni P, Peruzzini M, Polyhedron 6 (1987) 351.

Elschenbroich Ch, Salzer A, Organometallchemie, 2nd ed, B G Teubner, Stuttgart, 1988.

Fenske D, Maczek B, Maczek K Z Anorg Allg Chem 623 (1997) 1113.

Fermin M C, Stephan D W, J Am Chem Soc 117 (1995)

Fermin M C, Ho J, Stephan D W, J Am Chem Soc 116 (1994) 6033.

Fermin M C, Ho J, Stephan D W, Organometallics 14 (1995) 4247.

Garcia M E, Riera V, Ruiz M A, Rueda M T, Saez D, Organometallics 21 (2002) 5515.

Garrou P E, Chem Rev 81 (1981) 229.

Ginsberg A P, Lindsell W E, McCullough K J, Sprinkle C R, Welch A J, J Am Chem Soc 108 (1986) 403.

Gröer T, Baum G, Scheer M, Organometallics 17 (1998) 5916.

Heinicke J, Gupta N, Singh S, Surana A, Kühl O, Bansal R K, Karaghiosoff K, Vogt M, Z Anorg Allg Chem 628 (2002) 2869.

Hey-Hawkins E, Chem Rev 94 (1994) 1661.

Hey-Hawkins E, Kurz S, J Organomet Chem 479 (1994) 125.

Hey-Hawkins E, Kurz S, Sieler J, Baum G, J Organomet Chem 486 (1995) 229.

Hirth U-A, Malisch W, J Organomet Chem 439 (1992) C16.

Hitchcock P B, Lemos M A N D A, Meidine M F, Nixon J F, Pombeiro A J L, J Organomet Chem 402 (1991) C23.

Hitchcock P B, Maah M J, Nixon J F, Zora J A, Leigh G J, Bekar M A, Angew Chem Int Ed 26 (1987) 474.

Ho J, Rousseau R, Stephan D W, Organometallics 13 (1994) 1918.

Ho J, Stephan D W, Organometallics 10 (1991) 3005.

Ho J, Stephan D W, Organometallics 11 (1992) 1014.

Hoffmann R, Brücken zwischen Anorganischer und Organischer Chemie (Nobel Vortrag), Angew Chem 94 (1982) 725.

Hou Z, Breen T L, Stephan D W, Organometallics 12 (1993) 3158.

Huttner G, Borm J, Zsolnai L, J Organomet Chem 262 (1984) C33.

Huttner G, Borm J, Zsolnai L, J Organomet Chem 263 (1984) C33.

Huttner G, Mohr G, Friedrich P, Schmid H G, J Organomet Chem 160 (1978) 59

Huttner G, Weber U, Sigwarth B, Scheidsteger O, Lang H, Zsolnai L, J Organomet Chem 282 (1985) 331.

Kraus F, Korber N, Chem Eur J 11 (2005) 5945.

Krossing I, van Wüllen L, Chem Eur J 8 (2002) 700.

Kurz S, Hey-Hawkins E, J Organomet Chem 462 (1993) 203.

Leoni P, Marchetti F, Marchetti L, Passarelli V, Chem Comm (2004) 2346.

Leoni P, Pasquali M, Sommovigo M, Laschi F, Zanello P, Albinati A, Lianza F, Pregosin P S, Rügger H, Organometallics 12 (1993) 1702.

Leoni P, Pasquali M, Sommovigo M, Laschi F, Zanello P, Albinati A, Lianza F, Pregosin P S, Rügger H, Organometallics 13 (1994) 4017.

Lindsell W E, Chem Comm (1982) 1422.

Lorenz I-P, Pohl W, Nöth H, Schmidt M, J Organomet Chem 475 (1994) 211.

Maslennikov S V, Glueck D S, Yap G P A, Rheingold A L, Organometallics 15 (1996) 2483.

Ryan R C, Pittman C U Jr., O'Connor J P, Dahl L F, J Organomet Chem 193 (1980) 247.

Scherer O J, Sitzmann H, Wolmershäuser G, J Organomet Chem 268 (1984) C9.

Seyfarth D, Wood T G, Fackler J P Jr, Mazany A M, Organometallics 3 (1984) 1121.

Index